Surveys and Tutorials in the Applied Mathematical Sciences

Volume 14

Featuring short books of approximately 80-200pp, Surveys and Tutorials in the Applied Mathematical Sciences (STAMS) focuses on emerging topics, with an emphasis on emerging mathematical and computational techniques that are proving relevant in the physical, biological sciences and social sciences. STAMS also includes expository texts describing innovative applications or recent developments in more classical mathematical and computational methods.

This series is aimed at graduate students and researchers across the mathematical sciences. Contributions are intended to be accessible to a broad audience, featuring clear exposition, a lively tutorial style, and pointers to the literature for further study. In some cases a volume can serve as a preliminary version of a fuller and more comprehensive book.

Münevver Tezer-Sezgin • Canan Bozkaya

Boundary Element Method for Magnetohydrodynamic Flow

2D MHD Duct Flow Problems

Münevver Tezer-Sezgin
Department of Mathematics
Middle East Technical University
Ankara, Türkiye

Canan Bozkaya
Department of Mathematics
Middle East Technical University
Ankara, Türkiye

ISSN 2199-4765 ISSN 2199-4773 (electronic)
Surveys and Tutorials in the Applied Mathematical Sciences
ISBN 978-3-031-58352-0 ISBN 978-3-031-58353-7 (eBook)
https://doi.org/10.1007/978-3-031-58353-7

Mathematics Subject Classification: 65M38, 65N38, 76M15, 76W05, 76-10

This Springer imprint is published by the registered company Springer Nature Switzerland AG
The registered company address is: Gewerbestrasse 11, 6330 Cham, Switzerland

Paper in this product is recyclable.

To
Zeynep, Memnune, and Şerife, Ahmet

Preface

The content of this book has been the subject of research works which the authors have held over twenty-five years on the boundary element method (BEM) and magnetohydrodynamics (MHD). The BEM applications of some MHD flow problems are the result of their own research and reflected in the book by references to several publications of the authors. The idea of collecting them together in a book form is to draw the attention of the applied mathematicians and engineers in their researches to the powerful and efficient technique BEM for solving MHD flow problems. It introduces graduate students of applied mathematics, physics, and engineering to the fundamentals of MHD, to the theory of the boundary element method and the dual reciprocity BEM, and to the application of BEM to some benchmark MHD flow problems. These are MHD flow through ducts, MHD flow in infinite regions and between parallel infinite plates, MHD convection flow in enclosures, and inductionless MHD flow in cavities. The aim is for the graduate students and the researches to enhance their understanding and expertise in the numerical solutions of MHD duct flow problems using the boundary element method. A previous background or preliminary reading in either MHD or BEM could be an advantage. However, we derive the MHD flow equations, and apply the BEM showing the mathematical development in a deductive way that the book is largely self-contained.

We have correlated essential theory in magnetohydrodynamics and boundary element method, and we especially proceed to some notable applications with which we are most familiar from our research and experience. Certain fundamental topics are repeated briefly than elsewhere, because there are several well-written basic text books available on either magnetohydrodynamis or boundary element method, for example, by Müller and Bühler (*Magnetohydrodynamics in Channels and Containers*, 2001, Springer), Brebbia (*The Boundary Element Method for Engineers*, 1978, Pentech Press), and Patridge, Brebbia, and Wrobel (*The Dual Reciprocity Boundary Element Method*, 1992, Computational Mechanics Pub.). Our orientation in the application of MHD flow is liquids (incompressible fluid), and BEM or dual reciprocity BEM (DRBEM) solutions of MHD flow in ducts or in infinite regions. Special interest is given to partly insulated partly perfectly conducting boundaries, and the thickness of parabolic boundary layer emanating from the point of discontinuities is computed. Treatment of infinite regions with BEM is deeply explained by proving the convergence of infinite boundary integrals.

Magnetohydrodynamics describes the area combining classical fluid mechanics and electrodynamics. Thus, it deals with flows of electrically conducting fluids which are subject to a magnetic field and/or electric current driven by an external voltage. MHD has a wide range of applications in geophysics, astrophysics, industry, technology, and bio-medicine. MHD effects in technological applications are seen on working principles of technical devices and industrial production processes in terms of optimization and control. Among these, the liquid metal duct flow under the influence of a strong magnetic field is an active discipline in natural science and engineering. In this book, we concentrate on the numerical solutions of MHD flow in pipes (channels) with rectangular or circular cross-sections (ducts, cavities) or in infinite regions where the fluid is liquid metal. Temperature variation of the fluid in enclosures is also taken into account in some problems.

The boundary element method is a particular application of weighted residual techniques, and in that sense, it is related to finite element method (FEM). The BEM requires the fundamental solution to the governing equation of the problem which is taken as Poisson type equation in this book. Thus, the logarithmic function in terms of the distance between two points is the right function. The discretization is performed by subdividing the boundary of the region into a series of elements, and thus obtaining the problem solution in terms of boundary values only, with considerable computer time and memory savings. The nonhomogeneous and nonlinear terms are tackled with its generalized form, the dual reciprocity boundary element method, which eliminates the need of computation of domain integrals. The BEM and DRBEM are also well suited to problems defined in infinite domains such as those frequently occurring in stress analysis, and MHD flows in infinite regions and between infinite parallel plates for which the domain type methods like the FEM or the FDM are unsuitable. MHD flow equations are convection-diffusion type and coupled in terms of the velocity of the fluid and induced magnetic field. The fundamental solution to these coupled equations, derived by the authors in terms of modified Bessel functions by using the adjoint operator, is included in the book. The BEM procedure is applied in Chap. 4 of this book for a large number of MHD flow problems with the fundamental solution of coupled MHD equations derived in Chap. 3. Chapter 5 presents the DRBEM applications to the problems of MHD flow and MHD convection flow with or without induced magnetic field involving time derivatives, nonlinear convective terms, and Buoyancy terms, which all result in nonhomogeneity containing the main solution of the problem. The DRBEM is the unique technique for solving such problems using the boundary-only nature of the BEM and giving the solution at the same time in the interior of the domain.

The book consists of five chapters and is organized as follows. In Chap. 1, we review some concepts of electromagnetism and fluid dynamics from [1] giving the governing equations of MHD flow toward incompressible fluids in pipes. The mathematical models (partial differential equations) for engineering applications are formulated, namely: pressure-driven MHD duct flow, electrically driven MHD flow in the upper half plane and between parallel infinite plates, inductionless MHD flow with electric potential when magnetic field applies perpendicular or parallel to the pipe-axis, natural and mixed convection MHD flows in enclosures. Chapter 2 deals with the presentation of BEM for some solution of potential flow problems, especially Poisson's type equations together with the derivation of required fundamental solution as is in [2]. The nonlinearities, convective terms and time derivatives are treated with the DRBEM which uses the fundamental solution of simpler form of the differential equation (here, fundamental solution of Laplace equation) similar to [3]. In Chap. 3, the fundamental solution which applies the steady MHD flow equations in their original coupled form is derived. The derivation belongs to the author's previous research and published in their cited references. The application of the BEM and discretization procedure are also given together with the insertion of boundary conditions. In Chap. 4, we present numerical solutions of steady MHD duct flow problems using BEM with the fundamental solution obtained in Chap. 3 for coupled MHD equations. Pressure-driven or electrically driven MHD flows in rectangular or circular ducts containing partly insulated partly perfectly conducting boundaries, or in infinite regions, are simulated for increasing values of Hartmann number. Convergence of infinite boundary integrals resulting in infinite Hartmann walls and on the upper half plane is shown. The thickness of parabolic boundary layers emanating from the points of singularities on the partly insulated partly perfectly conducting boundaries is computed. The last chapter, Chap. 5, first gives DRBEM solution of unsteady MHD flow in a square cavity (Re and R_m are not neglected). Then, the DRBEM solutions are obtained and simulated for the MHD flow with heat transfer in a lid-driven cavity in terms of stream function-vorticity-temperature. In some application problems, magnetic potential and electric potential are also involved. Numerical results of inductionless flow are presented for Buoyancy-driven MHD flow, MHD flow with electric potential, and MHD flow under the magnetic field applied in the pipe-axis direction. Numerical results are deeply discussed for all the problem parameters for each considered MHD flow problems.

Ankara, Türkiye

Münevver Tezer-Sezgin
Canan Bozkaya

Contents

Acronyms

2D	two-dimensional
3D	three-dimensional
A	magnetic potential
$\mathbf{A}$	vector potential
$\mathbf{B}$	magnetic induction field
B	induced magnetic field (induced current)
BEM	boundary element method
(B_x, B_y)	induced magnetic field components
B_0	magnitude or intensity of applied magnetic field
c	wall conductance parameter, wall conductance ratio
c_A	coefficient of the BEM solution (V, B) at the point A
c_i	coefficient of the BEM solution u_i, $c_i = \theta_i/2\pi$
c_p	specific heat capacity
$\mathbf{D}$	electric induction
DRBEM	dual reciprocity boundary element method
$\mathbf{E}$	electric field
$\mathbf{E}'$	electric field in a reference frame
$\mathbf{f}$	force
$\mathbf{f}^{(em)}$	Lorentz force
$f(x, y)$	inhomogeneity in Poisson equation
$f(r)$	radial basis function
$\mathbf{F}$	coordinate matrix
g	gravitational acceleration constant
$\mathbf{g}$	gravitational acceleration vector
$\mathbf{G}$	BEM matrix
$\mathbf{G}^*$	fundamental solution for $\mathbf{L}^*$
$\mathbf{H}$	magnetic field, BEM matrix
Ha	Hartmann number, $Ha = \sqrt{{Ha_x}^2 + {Ha_y}^2}$
j	current density
$\mathbf{J}$	electric current density vector

k	induced current value at the walls
K_0	modified Bessel function of the second kind and of order zero
K_1	modified Bessel function of the second kind and of order one
ℓ	length of conducting portion
l	length of boundary element
L	number of interior points
$\mathbf{L}$	matrix differential operator for coupled MHD equations
$\mathbf{L}^*$	matrix adjoint operator of $\mathbf{L}$
L_0	characteristic length
$\mathbf{M}$	vector with components (Ha_x, Ha_y)
MHD	magnetohydrodynamics
$\mathbf{n}$	unit normal pointing outwards, $\mathbf{n} = (n_x, n_y)$
N	Stuart number, number of boundary elements
$N_1(\xi), N_2(\xi)$	linear shape functions
p	pressure of the fluid
Pr	Prandtl number
q	electric charge density
q^*	normal derivative of fundamental solution
Q	integrated flow rate, point charge, source of energy
r	distance
$\mathbf{r}$	distance vector, $\mathbf{r} = (r_x, r_y)$
R	residual, radius of infinitely distant semi-circle
Ra	Rayleigh number
Re	Reynolds number
R_m	magnetic Reynolds number
s	ratio of the pressure gradients of two coupled ducts
$\boldsymbol{\iota}$	unit tangent
T	temperature
T_c	temperature of cold wall
T_h	temperature of hot wall
$\mathbf{U}$	velocity field
U_0, V_0	characteristic velocity
u_z	pipe-axis velocity
u^*	fundamental solution
$\hat{u}, \hat{q}$	particular solution and its normal derivative
u^h	homogeneous solution
V	velocity
w	vorticity
$\mathbf{w}$	vector weight function, $\mathbf{w} = (V^*, B^*)$
α	slip length
$\boldsymbol{\alpha}$	vector containing coefficients α_j in the right-hand-side vector $\mathbf{b}$
$\bar{\alpha}$	thermal diffusivity
β	thermal expansion coefficient
$\boldsymbol{\beta}$	vector containing coefficients β_j in the approximation of solution u
Δ_i, Δ_A	Dirac delta function

Δt	time increment
ΔT	temperature difference
ϵ	the permittivity of the conducting material
η	magnetic diffusivity
γ	angle made with the x-axis or y-axis
Γ	boundary or interface
Γ_x	diameter of the semi-circle on the x-axis
Γ_∞	infinitely distant upper semi-circle
κ	thermal conductivity
κ_f	thermal conductivity of the fluid
μ	coefficient of viscosity or dynamic viscosity
μ_e	magnetic permeability
ν	kinematic viscosity
∇p	presure gradient
Ω	domain or finite volume
ϕ	electric potential
ϕ_w	wall potential
Φ	fundamental solution of biharmonic equation, kinetic energy
ψ	stream function
ρ	density of the fluid
ρ_0	density of the fluid at reference temperature
σ	electrical conductivity
θ_i	internal angle at the point i
θ_u, θ_q	relaxation parameters for u and q

List of Figures

List of Tables

Chapter 1
The Equations of Magnetohydrodynamics

1.1 Introduction

Magnetohydrodynamics (MHD) is the science or study of motion of electrically conducting fluids in the presence of magnetic fields. The fluid motion is influenced from the magnetic field, and this influence is expressed mathematically by including the electromagnetic force in the equations of motion. Hartmann and Lazarus [4] pointed out the influence of external magnetic field on the fluid motion by the experiments. The electromagnetic force causes an electric current density to flow in the fluid. Then, the fluid motion changes in turn the magnetic field through Ohm's law determining the appearance within the fluid of an induced magnetic field. Thus, the fluid mechanics equations from hydrodynamics and the electromagnetic field equations from electromagnetic phenomena must be solved simultaneously in terms of the velocity of the fluid and the induced magnetic field. This makes most MHD problems both theoretically and practically difficult to study since the governing equations are coupled.

MHD has important practical applications in such fields as astrophysics, geophysics, space propulsion, and electrical power generation. So, the applications of MHD may be grouped as [1]:

(i) Plasma confinement for the fusion reaction, propulsion, and flight control for rocket and hypersonic aerodynamic vehicles
(ii) Pumping of conducting liquids and power generation

Plasma represents the medium that is compressible, continuous (or discrete), and electrically conducting. Most of the matter in the universe is in the plasma state, e.g., interstellar, stellar, and solar matter. The cosmic flight of a satellite represents a motion of a body through a conducting medium in the presence of a magnetic field. In the hypersonic velocity aerodynamics, we have a magneto-aerodynamic problem since the gas in front of the moving body is heated by

M. Tezer-Sezgin, C. Bozkaya, *Boundary Element Method for Magnetohydrodynamic Flow*, Surveys and Tutorials in the Applied Mathematical Sciences 14,
https://doi.org/10.1007/978-3-031-58353-7_1

friction and becomes conductor. Concerning the possible engineering applications of magnetohydrodynamics, we may mention the flowmeter, the MHD generator, and the electromagnetic pump. In a flowmeter, a conducting fluid passes down an insulating pipe across which a steady magnetic field is applied and potential gradient is created and can be measured by probes embedded in the walls of the pipe (i.e., technique used to measure the flow of blood). MHD power generation is the conversion of kinetic or potential energy of a fluid with a magnetic field into electrical energy. The electromagnetic pump is a device that is the converse of MHD generators. A conducting fluid in a pipe is forced to move by a Lorentz force created when mutually perpendicular magnetic fields and electric currents are applied perpendicular to the pipe. The electromagnetic pumps allow a sensitive control of flow rate.

The design and construction of a liquid metal cooled fusion blanket, MHD generators, pumps, and accelerators require an understanding of the flow of a conducting fluid down a straight channel (pipe) of rectangular or circular cross section (duct) under an applied transverse magnetic field. The magnetic field is imposed either parallel to one pair of sides of the rectangular duct or with an angle made with the duct plane. There are some MHD flow problems also when the external magnetic field applies in the pipe-axis direction.

In this book, the problems of MHD flow in pipes are studied numerically, in particular the cases in which the magnitude and structure of the fluid velocity and induced magnetic field are dependent on the conductivities (insulated to perfectly conducting) and the slip (no-slip to slipping) of the duct walls. Special emphasis is given on the ducts that have partially conducting walls (partly insulated and partly perfectly conducting). Some problems of the flow of a viscous fluid, through a parallel-wall duct (flow between parallel plates) and on the half plane, are also studied with the emphasis on the mixed-conductivity boundaries. We are concerned only with the incompressible fluids and with the solutions of both steady and time-dependent two-dimensional MHD flow equations in either rectangular or circular ducts with the most general form of wall conditions. Also, the natural and mixed convection flows in enclosures in the presence of magnetic field are solved where the temperature equation is involved. Inductionless MHD flows in channels with or without electric potential and inductionless MHD flow under the pipe-axis direction magnetic field are considered too.

1.2 The Governing Equations of Electrodynamics and Fluid Mechanics

We shall present the governing equations of MHD flow from the Navier–Stokes equations and a simplified version of Maxwell's equations through Ohm's law based on the derivations given in [1]. A summary of electromagnetic and fluid mechanics equations from a physics and mathematical perspective will be provided for the case of incompressible fluid flows [1, 5–7]. The mathematical models (differential equations) for engineering applications such as pressure driven MHD duct flows, electrically driven MHD flow between parallel infinite plates, MHD flow in infinite space, MHD convection flows in enclosures, and inductionless MHD flow with electric potential under the magnetic field that is perpendicular or parallel to the pipe axis are all derived.

1.2.1 The Equations of Electromagnetism

The word magnetohydrodynamics covers the phenomena from electrodynamics and fluid dynamics in which the velocity **U** of an electrically conducting fluid couples the magnetic field **B**. In this book, centered on MHD coupling, the fluid is assumed to be incompressible, nonmagnetic, and its physical properties such as the magnetic permeability μ_e, the electrical conductivity σ, and the kinematic viscosity ν are assumed to be constant. It is known that liquid metals match these conditions best.

The fundamental notion of electricity is that of electric charge density q that plays a role analogous to that of mass in mechanics. The equation for the conservation of charge with the electric current density **J** is given as [1]

$$\frac{\partial q}{\partial t} + \operatorname{div} \mathbf{J} = 0, \tag{1.1}$$

where $\operatorname{div} \mathbf{J} = \nabla \cdot \mathbf{J}$ with the gradient operator ∇.

When the material is sufficiently conducting in order that the electric charge density becomes negligible as in the case of quasi-neutrality in plasma, the equation for the conservation of charge becomes [7]

$$\operatorname{div} \mathbf{J} = 0 \tag{1.2}$$

implying that in conducting materials charges are not accumulated.

1.2.1.1 Maxwell's Equations

Coulomb's law interprets the force, experienced by the charge Q to an electric field **E**, as [1, 7]

$$\mathbf{f} = Q\mathbf{E}, \tag{1.3}$$

where **E** is the electrical field created by summation of all charge densities q_i in space that are differently charged than the point charge Q.

Since electric charges are the origin of electric fields, there is a relation between the electric field and the spatial distribution of charges as

$$\nabla \cdot (\epsilon \mathbf{E}) = q, \quad \mathbf{D} = \epsilon \mathbf{E}, \tag{1.4}$$

where $\mathbf{D}$ is the electric induction, ϵ is the permittivity of the material, and q is the electric charge density. In electric conductors at steady conditions, localized charges are very small.

One of the origins of magnetic fields $\mathbf{H}$ is the electric current flow in conductors. Ampere's law states that the curl of magnetic field (curl $\mathbf{H} = \nabla \times \mathbf{H}$) is related to the electric current density $\mathbf{J}$ with

$$\nabla \times \mathbf{H} = \mathbf{J} + \frac{\partial}{\partial t}(\epsilon \mathbf{E}) \tag{1.5}$$

for fast electromagnetic processes such as the propagation of electromagnetic waves. Here, $\partial \mathbf{D}/\partial t = \partial(\epsilon \mathbf{E})/\partial t$ describes the so-called displacement currents. This relationship (1.5) is reduced to

$$\nabla \times \mathbf{H} = \mathbf{J} \tag{1.6}$$

for slowly varying electromagnetic processes in which the occurring velocities are much smaller compared to the speed of light. That is, displacement currents can be neglected.

Taking the divergence of (1.6), the left-hand side vanishes identically, and one obtains

$$\nabla \cdot \mathbf{J} = 0,$$

which is in accordance with Eq. (1.2).

Another basic observation in electrodynamics is the induction of a magnetic field in a conductor. The Faraday's law in differential form

$$\nabla \times \mathbf{E} = -\frac{\partial}{\partial t}(\mu_e \mathbf{H}) \tag{1.7}$$

with the definition of magnetic induction $\mathbf{B} = \mu_e \mathbf{H}$ takes the form

$$\nabla \times \mathbf{E} = -\frac{\partial \mathbf{B}}{\partial t}. \tag{1.8}$$

Taking the divergence of (1.8) again results in a constant $\nabla \cdot \mathbf{B}$ in time. However, the integration constant must be zero, since at an initial time, all currents are vanishing. Thus, generally, the magnetic induction field is solenoidal

$$\nabla \cdot \mathbf{B} = 0 \tag{1.9}$$

meaning that magnetic field lines are closed. Thus, Eqs. (1.4), (1.5) and (1.8), (1.9) relating the fields $\mathbf{E}$, $\mathbf{B}$, $\mathbf{D}$, $\mathbf{H}$, $\mathbf{J}$ and the electric charge density q, independently of the properties of the conductor matter, constitute the Maxwell's equations. These are [1]

$$
\begin{aligned}
&\operatorname{div} \mathbf{B} = 0 \\
&\operatorname{curl} \mathbf{E} = -\frac{\partial \mathbf{B}}{\partial t} \\
&\operatorname{div} \mathbf{D} = q \\
&\operatorname{curl} \mathbf{H} = \mathbf{J} + \frac{\partial \mathbf{D}}{\partial t}
\end{aligned}
\tag{1.10}
$$

in which the last equation becomes $\operatorname{curl} \mathbf{H} = \mathbf{J}$ when the displacement currents are neglected.

1.2.1.2 Lorentz Force and Ohm's Law

Another essential observation of the interaction between electric currents and magnetic fields is the action of forces. A given magnetic field acting on a conductor element, with current flow given by the flux of the current density **J**, describes the Lorentz force as [1]

$$\mathbf{f} = \mathbf{J} \times \mathbf{B}\,. \tag{1.11}$$

This force actually contains the Coulomb force $q\mathbf{E}$ as

$$\mathbf{f} = q\mathbf{E} + \mathbf{J} \times \mathbf{B}, \tag{1.12}$$

but it is neglected for the conducting materials with very small electric charge density q.

1.2.1.3 Ohm's Law

This constitutive law describes the linear relation

$$\mathbf{J} = \sigma \mathbf{E}' \tag{1.13}$$

relating the electric current density **J** with $\mathbf{E}'$ that measures the electric field in a frame of reference fixed to the material. It characterizes the ability of isotropic conducting materials (such as liquids) to transport electric charge under the influence of an applied electric field. In moving fluids, this property of material is still true for an observer attached to the particle. Thus, in the laboratory frame, for a fluid moving with a velocity $\mathbf{U}$

$$\mathbf{E}' = \mathbf{E} + \mathbf{U} \times \mathbf{B} \tag{1.14}$$

and including the transport of electric charge by convection, the electric current density takes the form (page 28 in [7])

$$\mathbf{J} = q\mathbf{U} + \sigma(\mathbf{E} + \mathbf{U} \times \mathbf{B}) \,. \tag{1.15}$$

However, in the electromagnetic approximation, the transport of electric charge by convection is negligible in comparison with transport by conduction proportional to σ. Thus, for moving conducting fluids, we have

$$\mathbf{J} = \sigma(\mathbf{E} + \mathbf{U} \times \mathbf{B}), \tag{1.16}$$

which is the expression for Ohm's law.

1.2.1.4 The Magnetic Induction Equation

The transport equation for the magnetic field $\mathbf{B}$ forms the basis of incompressible MHD theory [1, 7].

We take the curl of Ohm's law for moving fluids

$$\nabla \times \mathbf{J} = \sigma(\nabla \times \mathbf{E} + \nabla \times \mathbf{U} \times \mathbf{B}), \tag{1.17}$$

and using relations (1.6) and (1.8), we get

$$\frac{1}{\mu_e}\nabla \times \nabla \times \mathbf{B} = \sigma(-\frac{\partial \mathbf{B}}{\partial t} + \nabla \times \mathbf{U} \times \mathbf{B}) \,. \tag{1.18}$$

Using the vector identity

$$\nabla \times \mathbf{U} \times \mathbf{B} = (\mathbf{B} \cdot \nabla)\mathbf{U} - (\mathbf{U} \cdot \nabla)\mathbf{B} + \mathbf{U}\,\mathrm{div}\,\mathbf{B} - \mathbf{B}\,\mathrm{div}\,\mathbf{U} \tag{1.19}$$

and noting that $\mathrm{div}\,\mathbf{U} = 0$ and $\mathrm{div}\,\mathbf{B} = 0$ since μ_e is constant and the fluid is incompressible, the magnetic induction equation (1.18) becomes

$$\frac{\partial \mathbf{B}}{\partial t} + (\mathbf{U} \cdot \nabla)\mathbf{B} = \frac{1}{\mu_e \sigma}\nabla^2 \mathbf{B} + (\mathbf{B} \cdot \nabla)\mathbf{U} \tag{1.20}$$

since $\nabla \times \nabla \times \mathbf{B} = -\nabla^2 \mathbf{B}$. This equation describes the temporal evolution $\partial \mathbf{B}/\partial t$ of the magnetic field $\mathbf{B}$ due to convection $(\mathbf{U} \cdot \nabla)\mathbf{B}$, diffusion $\nabla^2 \mathbf{B}$, and magnetic field intensity sources $(\mathbf{B} \cdot \nabla)\mathbf{U}$ by the velocity field $\mathbf{U}$. The quantity $(\mu_e \sigma)^{-1} = \eta$ is the magnetic diffusivity.

Electromagnetic conditions at interfaces between two materials (usually between the fluid and the rigid wall or between two immiscible liquids) are derived from integral forms of Ampere's and Faraday's laws for a control volume enclosing the interface. Then, shrinking this control volume toward the interface, the conditions for the magnetic field, electric field, and current density are obtained [1]. These are $(\mathbf{B_1}-\mathbf{B_2})\cdot\mathbf{n}=0$, $(\mathbf{E_1}-\mathbf{E_2})\cdot\mathbf{t}=0$, $(\mathbf{J_1}-\mathbf{J_2})\cdot\mathbf{n}=0$, $(\mathbf{H_1}-\mathbf{H_2})\cdot\mathbf{t}=0$, where the subscripts 1 and 2 indicate the values on both sides of the interface, and $\mathbf{n}$ is the unit normal pointing from material 2 into material 1. Dot products with $\mathbf{n}$ imply that the normal components of the electromagnetic variables $\mathbf{B}$ and $\mathbf{J}$ are continuous across the interface, whereas dot product with $\mathbf{t}$ implies that the tangential components of the electric field and the magnetic field are continuous across the interface.

1.2.2 *Fluid Dynamics Equations*

Next, we will briefly give the balances for mass and momentum. We consider incompressible fluids such as liquid metals. The conservation of mass for incompressible fluids is then [1, 7]

$$\nabla\cdot\mathbf{U}=0\,. \tag{1.21}$$

The balance of linear momentum is written as (Navier–Stokes equations) [6]

$$\rho(\frac{\partial\mathbf{U}}{\partial t}+(\mathbf{U}\cdot\nabla)\mathbf{U})=-\nabla p+\rho\nu\nabla^2\mathbf{U}+\mathbf{f}^{(em)}, \tag{1.22}$$

where p denotes the pressure, ρ the fluid density, and $\nu=\mu/\rho$ is the kinematic viscosity with coefficient of viscosity (dynamic viscosity) μ. From Eq. (1.11)

$$\mathbf{f}^{(em)}=\mathbf{J}\times\mathbf{B} \tag{1.23}$$

is the Lorentz force neglecting Coulomb force $q\mathbf{E}$ in Eq. (1.12).

Equation (1.22) describes the temporal evolution of the linear momentum of a fluid element that changes by the action of pressure force $-\nabla p$, viscous friction $\rho\nu\nabla^2\mathbf{U}$, and by the Lorentz force, $\mathbf{J}\times\mathbf{B}$, when the given medium is simple. Thus, Lorentz force couples the mechanical and electrodynamic states of the fluid system. From Eqs. (1.6) and (1.11), we rewrite equation (1.22) neglecting the displacement currents

$$\rho(\frac{\partial\mathbf{U}}{\partial t}+(\mathbf{U}\cdot\nabla)\mathbf{U})=-\nabla p+\rho\nu\nabla^2\mathbf{U}+\frac{1}{\mu_e}(\nabla\times\mathbf{B})\times\mathbf{B}\,. \tag{1.24}$$

Since

$$(\nabla \times \mathbf{B}) \times \mathbf{B} = (\mathbf{B} \cdot \nabla)\mathbf{B} - \frac{1}{2}\nabla|\mathbf{B}|^2, \tag{1.25}$$

where $|\mathbf{B}|$ denotes the magnitude of $\mathbf{B}$. Equation (1.24) becomes then

$$\rho(\frac{\partial \mathbf{U}}{\partial t} + (\mathbf{U} \cdot \nabla)\mathbf{U}) = -\nabla p + \rho\nu\nabla^2\mathbf{U} + \frac{1}{\mu_e}(\mathbf{B} \cdot \nabla)\mathbf{B} - \frac{1}{2\mu_e}\nabla|\mathbf{B}|^2 . \tag{1.26}$$

Now, Eqs. (1.20) and (1.26) constitute the coupled transient MHD flow equations in terms of the velocity $\mathbf{U}$ of the fluid and the induced magnetic field $\mathbf{B}$.

In many engineering problems, the fluid is confined in a finite domain Ω bounded by an interface Γ by rigid walls (as in the cross section of a pipe (duct)). The flow starts with a constant pressure gradient, and it is fully developed, that is, the fluid variables do not change along the flow path and the derivatives of the velocities along the flow direction vanish. Kinematic constraint between the fluid and the rigid walls is the no-slip condition for the velocity since viscous fluids stick at rigid nonmoving walls ($\mathbf{U} = \mathbf{0}$ at Γ). For some rigid wall materials, it is possible for the fluid to slip in the vicinity of the wall that is called slip velocity. In this case, velocity of the fluid and its normal derivative are related with a slip length α as $\alpha\partial\mathbf{U}/\partial\mathbf{n} + \mathbf{U} = \mathbf{0}$.

1.2.3 Temperature Equation

The balance of the total energy in a finite volume Ω containing fluid and bounded by an interface Γ by rigid walls is a convection–diffusion equation for the temperature T of the form [1]

$$\rho c_p(\frac{\partial T}{\partial t} + (\mathbf{U} \cdot \nabla)T) = \nabla \cdot (\kappa\nabla T) + \frac{1}{\sigma}(\mathbf{J} \cdot \mathbf{J}) + \Phi + Q, \tag{1.27}$$

where the internal thermal energy is expressed by the temperature and material parameter that is the heat capacity ρc_p (c_p is the specific heat capacity). κ is the thermal conductivity, and when it is constant, $\kappa/\rho c_p = \bar{\alpha}$ denotes the thermal diffusivity. Equation (1.27) shows that the temporal increase (Ohmic heating) of the internal thermal energy $\rho c_p\, \partial T/\partial t$ equals to the loss of magnetic energy due to Joule dissipation $\frac{1}{\sigma}(\mathbf{J} \cdot \mathbf{J})$ plus the loss of kinetic energy Φ due to viscous dissipation.[1] This is an energy transfer mechanism in which magnetic energy is converted into thermal energy. Q is the other source of energy release due to chemical reactions, nuclear radiation, etc. The total energy should be conserved. For applications in liquid metal MHD channel flows, the last three terms due to Joule dissipation, viscous dissipation, and other sources can be neglected. This specific

[1] The irreversible conversion of work done against viscous forces into internal (thermal) energy.

form of the energy equation governs the temperature distribution in incompressible Newtonian fluids.

For the heat transfer duct flow problems, the reasonable physical condition at the interface Γ is the continuity of the heat flux ($\nabla T \cdot \mathbf{n} = 0$ at Γ). For the thin-wall assumption, thermal boundary conditions may be in terms of temperature, heat flux, or a combination of both at Γ.

1.2.4 Natural and Mixed Convection in the Presence of a Magnetic Field

The driving force for natural convection in a fluid flow is the temperature variation of the fluid in the region considered. When the fluid is heated, density decreases and the fluid rises. In a gravitational field, the net force between this movement and the gravitational force emerges, which is referred to a buoyancy force.

When the fluid motion is generated by an external force such as a pump, fan, or lid as well as the temperature difference on the walls of the region, it is called forced convection. Mixed convection flow is associated with both natural and forced convection flows according to the dominance of external force or temperature difference.

The two-dimensional mixed convection flow of an incompressible fluid is considered neglecting viscous dissipation and thermal radiation.[2]

In the presence of heat transfer and induced magnetic field, momentum equations will include buoyancy force as well as the Lorentz force. Assuming constant fluid properties (μ_e, σ, ν) except density variation in the buoyancy term, the buoyancy force $\mathbf{g}\beta(T - T_c)$ is added to the y-component of momentum equations (assuming $+y$ direction of gravitational acceleration vector $\mathbf{g}$) according to Boussinesq approximation that is [1]

$$\rho = \rho_0[1 - \beta(T - T_c)] \, .$$

Here, ρ_0 is the density of the fluid at the reference temperature T_c, β is the thermal expansion coefficient defined as $\beta = (-\frac{1}{\rho})(\frac{\partial \rho}{\partial T})_p$ at constant pressure p, T is the temperature of the fluid, and T_c is the cold wall temperature. The Hall effect[3] and Joule heating[4] effect are also neglected.

Then, the two-dimensional full MHD flow equations with heat transfer, which involve the continuity and the Navier–Stokes equations (1.21)–(1.22) including buoyancy force, and magnetic induction equations (1.20), are accompanied with the energy equation (1.27) and are given as [1]

---

[2] Electromagnetic radiation emitted by accelerated charged particles (due to heat) in matter.

[3] This effect is important for ionized gases in the case of strong magnetic field.

[4] Heat generated by the electric current passing through a resistor.

$$
\begin{aligned}
\nabla \cdot \mathbf{U} &= 0 \\
\nu \nabla^2 \mathbf{U} &= \frac{\partial \mathbf{U}}{\partial t} + \mathbf{U} \cdot \nabla \mathbf{U} + \frac{1}{\rho} \nabla p + \mathbf{g}\beta(T - T_c) - \frac{1}{\rho} \mathbf{J} \times \mathbf{B} \\
\nabla \cdot \mathbf{B} &= 0 \\
\frac{1}{\mu_e \sigma} \nabla^2 \mathbf{B} &= \frac{\partial \mathbf{B}}{\partial t} + (\mathbf{U} \cdot \nabla)\mathbf{B} - (\mathbf{B} \cdot \nabla)\mathbf{U} \\
\bar{\alpha} \nabla^2 T &= \frac{\partial T}{\partial t} + \mathbf{U} \cdot \nabla T
\end{aligned}
\tag{1.28}
$$

for an incompressible viscous fluid, where $\mathbf{U} = (u, v, 0)$ and $\mathbf{B} = (B_x, B_y, 0)$ and $\bar{\alpha} = \kappa_f / \rho c_p$ is the thermal diffusivity with thermal conductivity of the fluid κ_f and specific heat capacity c_p. This vector form of the equation may be rewritten explicitly as (since $\mathbf{J} \times \mathbf{B} = \frac{1}{\mu_e} (\nabla \times \mathbf{B}) \times \mathbf{B}$ from Eq. (1.6))

$$
\begin{aligned}
\frac{\partial u}{\partial x} + \frac{\partial v}{\partial x} &= 0 \\
\nu \nabla^2 u &= \frac{\partial u}{\partial t} + u \frac{\partial u}{\partial x} + v \frac{\partial u}{\partial y} + \frac{1}{\rho} \frac{\partial P}{\partial x} + \frac{B_y}{\rho \mu_e} (\frac{\partial B_y}{\partial x} - \frac{\partial B_x}{\partial y}) \\
\nu \nabla^2 v &= \frac{\partial v}{\partial t} + u \frac{\partial v}{\partial x} + v \frac{\partial v}{\partial y} + \frac{1}{\rho} \frac{\partial P}{\partial y} - \frac{B_x}{\rho \mu_e} (\frac{\partial B_y}{\partial x} - \frac{\partial B_x}{\partial y}) - g\beta(T - T_c) \\
\frac{1}{\sigma \mu_e} \nabla^2 B_x &= \frac{\partial B_x}{\partial t} + u \frac{\partial B_x}{\partial x} + v \frac{\partial B_x}{\partial y} - B_x \frac{\partial u}{\partial x} - B_y \frac{\partial u}{\partial y} \\
\frac{1}{\sigma \mu_e} \nabla^2 B_y &= \frac{\partial B_y}{\partial t} + u \frac{\partial B_y}{\partial x} + v \frac{\partial B_y}{\partial y} - B_x \frac{\partial v}{\partial x} - B_y \frac{\partial v}{\partial y} \\
\bar{\alpha} \nabla^2 T &= \frac{\partial T}{\partial t} + u \frac{\partial T}{\partial x} + v \frac{\partial T}{\partial y},
\end{aligned}
\tag{1.29}
$$

where g is the gravitational acceleration constant. Equations (1.29) are called the full MHD equations.

In two dimensions, it is sometimes useful to introduce the stream function and the vorticity form of Eq. (1.29). Defining the stream function $\psi(x, y)$ as $u = \frac{\partial \psi}{\partial y}$, $v = -\frac{\partial \psi}{\partial x}$ to satisfy the continuity condition $\nabla \cdot \mathbf{U} = 0$ in two dimensions, and the vorticity $\mathbf{w} = \nabla \times \mathbf{U} = (0, 0, w)$ with $w = \frac{\partial v}{\partial x} - \frac{\partial u}{\partial y}$, the momentum equations in (1.29) are transformed to stream function–vorticity form. That is, differentiating the x-momentum equation with respect to y and y-momentum equation with respect to x, and subtracting, pressure terms are eliminated, and the vorticity transport equation is obtained as

$$\begin{aligned}\nu\nabla^2 w &= \frac{\partial w}{\partial t} + u\frac{\partial w}{\partial x} + v\frac{\partial w}{\partial y} - g\beta\frac{\partial T}{\partial x} \\ &\quad - \frac{1}{\rho\mu_e}[B_x\frac{\partial}{\partial x}(\frac{\partial B_y}{\partial x} - \frac{\partial B_x}{\partial y}) + B_y\frac{\partial}{\partial y}(\frac{\partial B_y}{\partial x} - \frac{\partial B_x}{\partial y})]\,.\end{aligned} \tag{1.30}$$

Also, the z-component of vorticity $w = \frac{\partial v}{\partial x} - \frac{\partial u}{\partial y}$ gives the stream function equation

$$\nabla^2\psi = -w\,. \tag{1.31}$$

Nondimensionalization of Eqs. (1.30), (1.31), and magnetic field equations and temperature equation in (1.29) is performed by taking dimensionless variables as

$$\begin{gathered}x' = \frac{x}{L_0},\ y' = \frac{y}{L_0},\ u' = \frac{u}{U_0},\ v' = \frac{v}{U_0},\ t' = \frac{tU_0}{L_0},\ \psi' = \frac{\psi}{L_0U_0},\ w' = \frac{wL_0}{U_0}, \\ T' = \frac{T-T_c}{T_h-T_c},\ B_x' = \frac{B_x}{B_0},\ B_y' = \frac{B_y}{B_0}\end{gathered}, \tag{1.32}$$

where B_0 is the magnitude of the applied magnetic field and T_h is the temperature of the hot wall. U_0 and L_0 are the characteristic velocity and the characteristic length of the problem region, respectively.

Hence, the nondimensional form of full MHD equations with heat transfer in terms of stream function–vorticity (dropping the prime notation) becomes

$$\begin{aligned}\nabla^2\psi &= -w \\ \frac{1}{Re}\nabla^2 w &= \frac{\partial w}{\partial t} + u\frac{\partial w}{\partial x} + v\frac{\partial w}{\partial y} - \frac{Ra}{PrRe^2}\frac{\partial T}{\partial x} \\ &\quad - \frac{Ha^2}{ReR_m}[B_x\frac{\partial}{\partial x}(\frac{\partial B_y}{\partial x} - \frac{\partial B_x}{\partial y}) + B_y\frac{\partial}{\partial y}(\frac{\partial B_y}{\partial x} - \frac{\partial B_x}{\partial y})] \\ \frac{1}{R_m}\nabla^2 B_x &= \frac{\partial B_x}{\partial t} + u\frac{\partial B_x}{\partial x} + v\frac{\partial B_x}{\partial y} - B_x\frac{\partial u}{\partial x} - B_y\frac{\partial u}{\partial y} \\ \frac{1}{R_m}\nabla^2 B_y &= \frac{\partial B_y}{\partial t} + u\frac{\partial B_y}{\partial x} + v\frac{\partial B_y}{\partial y} - B_x\frac{\partial v}{\partial x} - B_y\frac{\partial v}{\partial y} \\ \frac{1}{PrRe}\nabla^2 T &= \frac{\partial T}{\partial t} + u\frac{\partial T}{\partial x} + v\frac{\partial T}{\partial y}\end{aligned}, \tag{1.33}$$

where

$$Re = \frac{U_0 L_0}{\nu},\ R_m = U_0 L_0 \sigma \mu_e,\ Ra = \frac{g\beta \Delta T L_0^3}{\nu \bar{\alpha}},\ Pr = \frac{\nu}{\bar{\alpha}},\ Ha = B_0 L_0 \frac{\sqrt{\sigma}}{\sqrt{\rho \nu}} \tag{1.34}$$

are the Reynolds number, magnetic Reynolds number, Rayleigh number, Prandtl number, and Hartmann number, respectively. Here, ΔT is the imposed temperature difference between the hot and cold walls.

In hydrodynamics, the Reynolds number represents the ratio of inertial to viscous forces. For a variety of applications in strong magnetic fields, the essential balance of forces establishes between electromagnetic forces and viscous forces. The ratio of the forces can be expressed in terms of the Reynolds number and the interaction parameter as $NRe = Ha^2$, where

$$N = \frac{\sigma L_0 B_0^2}{\rho U_0} \tag{1.35}$$

is called Stuart number. Hartmann number was first used by Hartmann [4] when dealing with the flow between parallel nonconducting planes.

The magnetic Reynolds number R_m represents the ratio of advection ($\nabla \times (\mathbf{U} \times \mathbf{B})$) to diffusion of the magnetic field ($\nabla^2 \mathbf{B}/\sigma \mu_e$). If $R_m >> 1$, advection dominates over diffusion, whereas diffusion dominates over advection if $R_m << 1$. In other words, $\mathbf{U}$ has little impact on $\mathbf{B}$ if $R_m << 1$. In this case, induced magnetic field is neglected, and the damping effect on fluid motion is only the applied magnetic field.

The Prandtl number Pr (the ratio of the kinematic viscosity and thermal conductivity) is a parameter that characterizes the fluid properties. Rayleigh number Ra shows a critical value that the transition between the static diffusive state and that of natural convection occurs.

Ampere's law (1.6) enables one to write the current density equation, i.e., $\mu_e \mathbf{J} = \nabla \times \mathbf{B}$, where $\mathbf{J} = (0, 0, j)$, in terms of induced magnetic field components, $\mathbf{B} = (B_x, B_y, 0)$ similar to vorticity definition as

$$j = \frac{1}{\mu_e}\left(\frac{\partial B_y}{\partial x} - \frac{\partial B_x}{\partial y}\right). \tag{1.36}$$

A Poisson type equation for the current density j may also be obtained by using Eq. (1.36). Differentiating the fourth and third equations of (1.29) with respect to x and y, respectively, then subtracting from each other, and using the divergence-free conditions $\nabla \cdot \mathbf{U} = 0$, $\nabla \cdot \mathbf{B} = 0$, the current density equation is deduced as

$$\begin{aligned}\frac{1}{\sigma}\nabla^2 j = \mu_e(\frac{\partial j}{\partial t} + u\frac{\partial j}{\partial x} + v\frac{\partial j}{\partial y}) - (B_x\frac{\partial w}{\partial x} + B_y\frac{\partial w}{\partial y}) \\ -2\left[\frac{\partial B_x}{\partial x}(\frac{\partial v}{\partial x} + \frac{\partial u}{\partial y}) + \frac{\partial v}{\partial y}(\frac{\partial B_x}{\partial y} + \frac{\partial B_y}{\partial x})\right].\end{aligned} \tag{1.37}$$

Defining the nondimensional variables in Eq. (1.32) and $j' = j/(B_0U_0\sigma)$, Eq. (1.37) is put into dimensionless form (dropping the prime notation)

$$\begin{aligned}\nabla^2 j = R_m(\frac{\partial j}{\partial t} + u\frac{\partial j}{\partial x} + v\frac{\partial j}{\partial y}) - (B_x\frac{\partial w}{\partial x} + B_y\frac{\partial w}{\partial y}) \\ -2\left[\frac{\partial B_x}{\partial x}(\frac{\partial v}{\partial x} + \frac{\partial u}{\partial y}) + \frac{\partial v}{\partial y}(\frac{\partial B_x}{\partial y} + \frac{\partial B_y}{\partial x})\right].\end{aligned} \tag{1.38}$$

Using the similar idea of satisfying the continuity equation, the solenoidal nature of **B** is also satisfied defining a vector potential **A** as $\mathbf{B} = \nabla \times \mathbf{A}$ in which $\mathbf{A} = (0, 0, A)$. In this fashion, $B_x = \partial A/\partial y$, $B_y = -\partial A/\partial x$.

Now, Eq. (1.36) becomes

$$j = \frac{1}{\mu_e}(\frac{\partial}{\partial x}(-\frac{\partial A}{\partial x}) - \frac{\partial}{\partial y}(\frac{\partial A}{\partial y})) \tag{1.39}$$

giving

$$\nabla^2 A = -\mu_e\, j, \tag{1.40}$$

which is similar to the relation $\nabla^2\psi = -w$ between stream function and vorticity.

The nondimensional variable $A' = A/(B_0L)$ provides us to get the nondimensional form of magnetic potential equation as

$$\nabla^2 A = -R_m\, j. \tag{1.41}$$

Analogous to the expressions of the momentum continuity equations in terms of stream function and vorticity, the magnetic induction equations are also written in terms of the magnetic potential and the current density. Thus, Buoyancy MHD flow with magnetic potential is described with the equations [8]

$$\begin{aligned}&\nabla^2\psi = -w \\ &\frac{1}{Re}\nabla^2 w = \frac{\partial w}{\partial t} + u\frac{\partial w}{\partial x} + v\frac{\partial w}{\partial y} - \frac{Ra}{Pr Re^2}\frac{\partial T}{\partial x} - \frac{Ha^2}{Re}(B_x\frac{\partial j}{\partial x} + B_y\frac{\partial j}{\partial y}) \\ &\nabla^2 A = -R_m\, j\end{aligned}$$

$$
\begin{aligned}
\frac{1}{R_m}\nabla^2 j &= \frac{\partial j}{\partial t} + u\frac{\partial j}{\partial x} + v\frac{\partial j}{\partial y} - \frac{1}{R_m}(B_x\frac{\partial w}{\partial x} + B_y\frac{\partial w}{\partial y}) \\
&\quad - \frac{2}{R_m}\left[\frac{\partial B_x}{\partial x}(\frac{\partial v}{\partial x} + \frac{\partial u}{\partial y}) + \frac{\partial v}{\partial y}(\frac{\partial B_x}{\partial y} + \frac{\partial B_y}{\partial x})\right] \\
\frac{1}{PrRe}\nabla^2 T &= \frac{\partial T}{\partial t} + u\frac{\partial T}{\partial x} + v\frac{\partial T}{\partial y}\,. \qquad (1.42)
\end{aligned}
$$

1.2.5 Inductionless MHD Convection Flow

If the magnetic Reynolds number is small, quasi-static approximation of the induction equation [9] as $R_m << 1$, the magnetic field induced by currents in the fluid is negligible compared to the externally applied field $\mathbf{B}$. In that situation, the flow no longer affects the magnetic field. Assuming a vertically applied constant magnetic field $\mathbf{B} = (0, B_0, 0)$ and the velocity field $\mathbf{U} = (u, v, 0)$, we have

$$
\mathbf{U}\times\mathbf{B} = \begin{vmatrix} \mathbf{i} & \mathbf{j} & \mathbf{k} \\ u & v & 0 \\ 0 & B_0 & 0 \end{vmatrix} = uB_0\mathbf{k} \qquad (1.43)
$$

and

$$
\mathbf{J}\times\mathbf{B} = \sigma(\mathbf{U}\times\mathbf{B})\times\mathbf{B} = \begin{vmatrix} \mathbf{i} & \mathbf{j} & \mathbf{k} \\ 0 & 0 & u\sigma B_0 \\ 0 & B_0 & 0 \end{vmatrix} = -\sigma u B_0^2\mathbf{i}\,. \qquad (1.44)
$$

Since Ohm's law $\mathbf{J} = \sigma(-\nabla\phi + \mathbf{U}\times\mathbf{B})$ with $\mathbf{E} = -\nabla\phi$ (ϕ is the electric potential), the conservation of current density div $\mathbf{J} = 0$ is reduced to $\nabla^2\phi = 0$ of which has a unique solution $\nabla\phi = 0$ if the boundary of the region is considered electrically insulating. As a conclusion, the term $\sigma u B_0^2$ is seen in the x-component of momentum equation in (1.29) as [1]

$$
\begin{aligned}
&\frac{\partial u}{\partial x} + \frac{\partial v}{\partial y} = 0 \\
&\nu\nabla^2 u = \frac{\partial u}{\partial t} + u\frac{\partial u}{\partial x} + v\frac{\partial u}{\partial y} + \frac{1}{\rho}\frac{\partial p}{\partial x} + \frac{\sigma B_0^2}{\rho}u \\
&\nu\nabla^2 v = \frac{\partial v}{\partial t} + u\frac{\partial v}{\partial x} + v\frac{\partial v}{\partial y} + \frac{1}{\rho}\frac{\partial p}{\partial y} - g\beta(T - T_c) \qquad (1.45) \\
&\bar{\alpha}\nabla^2 T = \frac{\partial T}{\partial t} + u\frac{\partial T}{\partial x} + v\frac{\partial T}{\partial y}
\end{aligned}
$$

since induced magnetic field equations are dropped.

Then, the nondimensional governing equations in terms of stream function $\psi(x, y)$, temperature $T(x, y)$, and vorticity $w(x, y)$ using the same nondimensional variables as in Eqs. (1.32) are written as

$$\begin{aligned}
\nabla^2\psi &= -w \\
\frac{1}{Re}\nabla^2 w &= \frac{\partial w}{\partial t} + u\frac{\partial w}{\partial x} + v\frac{\partial w}{\partial y} - \frac{Ra}{PrRe^2}\frac{\partial T}{\partial x} - \frac{Ha^2}{Re}\frac{\partial u}{\partial y} \\
\frac{1}{PrRe}\nabla^2 T &= \frac{\partial T}{\partial t} + u\frac{\partial T}{\partial x} + v\frac{\partial T}{\partial y}\,.
\end{aligned} \tag{1.46}$$

When external magnetic field applies horizontally, $\mathbf{B} = (B_0, 0, 0)$, Eqs. (1.46) will contain the last term $\frac{Ha^2}{Re}\frac{\partial v}{\partial x}$ in the vorticity equation as the only change,

$$\begin{aligned}
\nabla^2\psi &= -w \\
\frac{1}{Re}\nabla^2 w &= \frac{\partial w}{\partial t} + u\frac{\partial w}{\partial x} + v\frac{\partial w}{\partial y} - \frac{Ra}{PrRe^2}\frac{\partial T}{\partial x} + \frac{Ha^2}{Re}\frac{\partial v}{\partial x} \\
\frac{1}{PrRe}\nabla^2 T &= \frac{\partial T}{\partial t} + u\frac{\partial T}{\partial x} + v\frac{\partial T}{\partial y}\,.
\end{aligned} \tag{1.47}$$

Equations (1.46) or (1.47) are going to be solved with proper boundary conditions. Stream function is a natural streamline on the boundaries, that is, it may be a constant or derived from its velocity relations if the velocity is specified on the boundary. Vorticity boundary condition is usually not known, and it is approximated by using the stream function equation $\nabla^2\psi = -w$. Temperature boundary conditions may be $T_{constant}$ according to hot and cold walls or $\partial T/\partial n = 0$ if the walls are adiabatic.

1.2.6 Inductionless MHD Flow with Electric Potential

In order to calculate MHD flows in piping systems where the duct walls are electrically conducting, it is often useful to work with the variables, velocity $\mathbf{U}$, and electric potential ϕ instead of using the velocity $\mathbf{U}$ together with the induced magnetic field $\mathbf{B}$. It is assumed that the fluid motion cannot change significantly the magnetic field; hence, curl $\mathbf{E} = \mathbf{0}$ and $\mathbf{E}$ is a gradient of a potential. This may happen when electrical conductivity or magnetic permeability of the fluid is low giving small magnetic Reynolds number. When the quasi-static approximation is assumed, the magnetic Reynolds number is small as $R_m << 1$ [9], the magnetic field induced by currents in the field is neglected.

The basic MHD equations then are the momentum equations with Lorentz force (Eq. (1.22)) and the Ohm's law

$$\mathbf{J} = \sigma(\mathbf{E} + \mathbf{U} \times \mathbf{B}),$$

where the electric field is expressed as the gradient of the scalar electric potential ϕ as $\mathbf{E} = -\nabla\phi$. Using the charge conservation equation div $\mathbf{J} = 0$, we have $\sigma(\nabla \cdot (-\nabla\phi) + \text{div}\,(\mathbf{U} \times \mathbf{B})) = 0$, and, therefore, the electric field potential equation becomes [1]

$$\nabla^2\phi = \text{div}\,(\mathbf{U} \times \mathbf{B})\,. \tag{1.48}$$

To accompany this electric potential equation (1.48) with momentum equations, we consider the steady, viscous flow of an incompressible, electrically conducting fluid in a channel with electrically conducting walls under an oblique magnetic field applied in plane (perpendicular to the flow direction). Then, the governing equations are from (1.22) and (1.23)

$$\begin{aligned} \rho(\mathbf{U} \cdot \nabla)\mathbf{U} &= -\nabla p + \rho\nu\nabla^2\mathbf{U} + \mathbf{J} \times \mathbf{B} \\ \nabla^2\phi &= \text{div}(\mathbf{U} \times \mathbf{B})\,. \end{aligned} \tag{1.49}$$

This set of Eqs. (1.49) has to be solved for adequate kinematic and electrical boundary conditions at the fluid–wall interface. The kinematic boundary condition is usually the no-slip velocity condition. For insulating surfaces, $\mathbf{J} \cdot \mathbf{n} = 0$, which yields a vanishing normal derivative of the potential $\mathbf{n} \cdot \nabla\phi = \frac{\partial\phi}{\partial n} = 0$ since $\mathbf{n} \cdot (\mathbf{U} \times \mathbf{B}) = 0$. For a perfectly conducting wall, the potential at the wall becomes uniform. The value of the wall potential then can be set to zero without loss of generality. For thin wall with finite conductivity, the local current entering the wall is discharged into the thin wall. This is described as $\frac{\partial\phi}{\partial n} + \nabla(c\nabla\phi_w)$=0 at the wall where c is the wall conductance parameter and ϕ_w stands for the wall potential.

The variation of the velocity is assumed to be only in the pipe-axis direction as $\mathbf{U} = (0, 0, u_z(x, y))$ for fully developed flow, and the external magnetic field $\mathbf{B} = (B_0 \cos\gamma, B_0 \sin\gamma, 0)$ is applied obliquely only in the plane (the cross section of the pipe) where γ is the angle made with the x-axis. Then, the zth component of Eqs. (1.49) gives (evaluating $\mathbf{J} \times \mathbf{B}$ for this form of $\mathbf{U}$ and $\mathbf{B}$) [10]

$$\begin{aligned} \rho\nu\nabla^2 u_z - B_0^2\sigma u_z &= \frac{\partial p}{\partial z} + \sigma B_0 \sin\gamma\frac{\partial\phi}{\partial x} - B_0\sigma\cos\gamma\frac{\partial\phi}{\partial y} \\ \nabla^2\phi &= -B_0\sin\gamma\frac{\partial u_z}{\partial x} + B_0\cos\gamma\frac{\partial u_z}{\partial y}\,. \end{aligned} \tag{1.50}$$

Nondimensionalization is performed by taking

$$x' = \frac{x}{L_0},\ y' = \frac{y}{L_0},\ u_z' = \frac{u_z}{U_0},\ \frac{\partial p}{\partial z} = -\frac{\rho \nu U_0}{L_0^2},\ \phi' = \frac{\phi}{\mu_e L_0 U_0 B_0} \tag{1.51}$$

to obtain dimensionless equations (dropping the prime notation)

$$\begin{aligned} \nabla^2 u_z - Ha^2 u_z &= -1 + Ha^2 (\sin\gamma \frac{\partial \phi}{\partial x} - \cos\gamma \frac{\partial \phi}{\partial y}) \\ \nabla^2 \phi &= -\sin\gamma \frac{\partial u_z}{\partial x} + \cos\gamma \frac{\partial u_z}{\partial y} \,. \end{aligned} \tag{1.52}$$

When the angle $\gamma = \pi/2$, the applied magnetic field is in the y-direction (vertically applied in the duct), whereas when $\gamma = 0$, it is horizontally applied. Equations (1.50) will simplify then according to the values of $\sin\gamma$ and $\cos\gamma$.

1.2.7 Inductionless MHD Flow Under the Magnetic Field in the Pipe-Axis Direction

The inductionless MHD flow in channels is also considered when the external magnetic field is applied in the direction of the flow (pipe-axis direction). This way electric potential is also generated due to the channel walls [1, 11].

The velocity has three components as $\mathbf{U} = (u_x, u_y, u_z)$ each varying in the duct xy-plane and $\mathbf{B} = (0, 0, B_0)$ influences the fluid in the streamwise direction. The flow is steady, and the fluid starts to move with a constant pressure gradient ∇p. The pressure is divided into cross-section pressure $p(x, y)$ and the pipe-axis pressure $p_z(z)$ with constant $\partial p_z / \partial z$.

Then, the two-dimensional equations (1.49) in the cross section (duct) of the channel are

$$\begin{aligned} \rho(u_x \frac{\partial u_x}{\partial x} + u_y \frac{\partial u_x}{\partial y}) - \rho\nu \nabla^2 u_x + \frac{\partial p}{\partial x} &= -\frac{\partial \phi}{\partial y} B_0 - B_0^2 u_x \\ \rho(u_x \frac{\partial u_y}{\partial x} + u_y \frac{\partial u_y}{\partial y}) - \rho\nu \nabla^2 u_y + \frac{\partial p}{\partial y} &= \frac{\partial \phi}{\partial x} B_0 - B_0^2 u_y \\ \rho(u_x \frac{\partial u_z}{\partial x} + u_y \frac{\partial u_z}{\partial y}) - \rho\nu \nabla^2 u_z &= -\frac{\partial p_z}{\partial z} \\ \nabla^2 \phi = B_0 (\frac{\partial u_y}{\partial x} - \frac{\partial u_x}{\partial y}) &= B_0 w, \end{aligned} \tag{1.53}$$

where $w = \frac{\partial u_y}{\partial x} - \frac{\partial u_x}{\partial y}$ is the vorticity.

Introducing stream function $\psi(x, y)$ with $\frac{\partial\psi}{\partial y} = u_x$, $\frac{\partial\psi}{\partial x} = -u_y$ and nondimensionalizing equation (1.53) using dimensionless variables (1.51), the dimensionless equations become

$$\begin{aligned}
\nabla^2\psi &= -w \\
\nabla^2\phi &= w \\
\frac{1}{N}(u_x\frac{\partial w}{\partial x} + u_y\frac{\partial w}{\partial y}) - \frac{1}{Ha^2}\nabla^2 w &= 0 \\
\frac{1}{N}(u_x\frac{\partial u_z}{\partial x} + u_y\frac{\partial u_z}{\partial y}) - \frac{1}{Ha^2}\nabla^2 u_z &= -\frac{1}{N}\frac{\partial p_z}{\partial z}\,.
\end{aligned} \tag{1.54}$$

Equations (1.54) are going to be solved iteratively starting from an initial estimate for vorticity, a given constant pressure gradient $\frac{\partial p_z}{\partial z}$. The boundary of the problem region is a natural streamline implying constant (or zero) values of stream function on the boundary and no-slip condition for u_z. Electric potential satisfies either $\frac{\partial\phi}{\partial n} = 0$ for insulating walls or $\frac{\partial\phi}{\partial n} + \nabla(c\nabla\phi_{wall}) = 0$ for variably conducting walls with wall conductance parameter c.

1.2.8 MHD Duct Flow Equations

When the flow is restricted to two dimensions as in the case of the cross section (duct) of the pipes or channels, the equations will be simplified further since both the velocity and the induced magnetic field will have components in the pipe-axis direction, and all the conditions stay invariant in that direction apart from a pressure gradient (fully developed). Hence, considering the pipe extension in the z-direction, the differentiation of the momentum equation (1.26) with respect to z shows that $\nabla(\frac{\partial p}{\partial z})$ vanishes. Thus, $\partial p/\partial z$ is a constant, and the MHD flow is reduced to the two-dimensional duct region in the xy-plane with the unidirectional velocity $\mathbf{U} = (V_x, V_y, V_z(x, y, t))$ and the magnetic field $\mathbf{B} = (B_x, B_y, B_z(x, y, t))$. The components of the magnetic field $B_x = B_0 \sin\gamma$ and $B_y = B_0 \cos\gamma$ are known in terms of the intensity B_0 and the direction of the externally applied magnetic field with an inclination angle γ made with the y-axis. Thus, the only unknowns remaining in the magnetic induction equation (1.20) and momentum equation (1.26) are the time-dependent pipe-axis velocity V_z and the induced magnetic field B_z varying only in the duct region.

With all the simplifications for the unsteady two-dimensional fully developed MHD flow, the governing equations take the forms [7, 12, 13] (since $V_x = 0$, $V_y = 0$, $B_x = B_0 \sin\gamma$, and $B_y = B_0 \cos\gamma$ constants, then z-component of $\mathbf{J}$ is $j = 0$, $\mathbf{U} \cdot \nabla = 0$ and $\mathbf{B} \cdot \nabla = B_0 \sin\gamma \frac{\partial}{\partial x} + B_0 \cos\gamma \frac{\partial}{\partial y}$.)

$$
\begin{aligned}
&\mu\nabla^2 V_z + \frac{1}{\mu_e}(B_0 \sin\gamma \frac{\partial B_z}{\partial x} + B_0 \cos\gamma \frac{\partial B_z}{\partial y}) = \frac{\partial p}{\partial z} + \rho\frac{\partial V_z}{\partial t} \\
&\frac{1}{\mu_e \sigma}\nabla^2 B_z + (B_0 \sin\gamma \frac{\partial V_z}{\partial x} + B_0 \cos\gamma \frac{\partial V_z}{\partial y}) = \frac{\partial B_z}{\partial t}\,.
\end{aligned} \tag{1.55}
$$

It is useful to transform Eqs. (1.55) into a dimensionless form by introducing scaled variables as follows [12]:

$$
V = \frac{V_z}{V_0},\; B = \frac{1}{V_0 \mu_e}(\sigma\mu)^{-1/2} B_z,\; x' = \frac{x}{L_0},\; y' = \frac{y}{L_0},\; t' = \frac{t V_0}{L_0}, \tag{1.56}
$$

where $V_0 = -L_0^2(\frac{\partial p}{\partial z})/\mu$ is the characteristic velocity (mean-axis velocity) and L_0 is the characteristic length. Substituting these scaled variables in Eq. (1.55) and dropping the prime notation, we obtain the following nondimensional equations for the time-dependent MHD duct flow in terms of the velocity $V(x, y, t)$ and the induced magnetic field $B(x, y, t)$:

$$
\begin{aligned}
\nabla^2 V + Ha_x \frac{\partial B}{\partial x} + Ha_y \frac{\partial B}{\partial y} &= -1 + Re\frac{\partial V}{\partial t} \\
\nabla^2 B + Ha_x \frac{\partial V}{\partial x} + Ha_y \frac{\partial V}{\partial y} &= R_m \frac{\partial B}{\partial t}\,.
\end{aligned} \tag{1.57}
$$

The dimensionless parameters Re, R_m, and Ha are the Reynolds number, magnetic Reynolds number, and Hartmann number ($Ha = \sqrt{Ha_x^2 + Ha_y^2}$ with $Ha_x = Ha \sin\gamma$ and $Ha_y = Ha\cos\gamma$), respectively, given in Eq. (1.34).

The duct walls on which the applied magnetic field has a normal component are called the Hartmann walls, whereas the walls tangential to the field are the side walls.

1.2.8.1 Boundary Conditions

We consider first viscous fluids with no-slip at the fluid wall interface Γ. Therefore all the velocity components vanish at the wall, and the hydrodynamic boundary condition reads $V = 0$ at the duct walls Γ. In some MHD duct flow problems, slip condition is also utilized for the velocity when the wall material allows the fluid to slip in the vicinity of the wall.

The electromagnetic boundary conditions are controlled by the electrical conductivity σ_{wall} of the duct wall material. If the walls are insulating, no current may enter an insulating wall from the fluid side, and as a consequence, the normal component of the current density vanishes at the wall, i.e., $\mathbf{J} \cdot \mathbf{n} = 0$ at Γ. The induced magnetic field B serves as a streamline for the current in the duct. If the wall is insulating, then the duct wall represents one of these isolines, and that lines of B being constant (current streamlines) may be taken as $B = 0$ on Γ, [1]. If the walls of the duct consist of a metallic material with finite conductivity, currents may cross the interface between the fluid and the wall. In the case of thin duct walls, i.e., $\frac{t_{\text{wall}}}{L_0} << 1$, t_{wall} being the wall thickness, the boundary condition for the induced magnetic field becomes for nonferromagnetic materials with $\mu_e = \mu_{e_{\text{wall}}}$,

$$\frac{\partial B}{\partial n} = -\frac{\sigma L_0}{\sigma_{\text{wall}} t_{\text{wall}}} \qquad \text{at } \Gamma.$$

This can be written as

$$\frac{\partial B}{\partial n} + cB = 0 \qquad \text{at } \Gamma \tag{1.58}$$

with the wall conductance ratio $c = \frac{\sigma L_0}{\sigma_{\text{wall}} t_{\text{wall}}}$ (the electrical conductance of the fluid divided by the electrical conductance of the wall material). Thus, the limiting cases occur when $\sigma_{\text{wall}} = 0$, $c \to \infty$ for insulating walls, that is, we have $B = 0$, and for perfectly conducting walls $\sigma_{\text{wall}} \to \infty$, $c = 0$, we have $\frac{\partial B}{\partial n} = 0$ at Γ, [1].

Now, the MHD duct flow equations (1.57) are going to be solved first for steady case and for several orientations of the applied magnetic field. These are two coupled second-order linear partial differential equations. For each problem considered in this book, the geometry and the boundary conditions will be explained when they are solved numerically in Chap. 4. Also, the characteristic length L_0 will be taken differently for each problem considered and specified according to the geometry of the problem. The characteristic velocity will be in general the mean-axis velocity of the pipe. For all the problems considered, the duct walls are a combination of insulators ($\sigma_{\text{wall}} \approx 0$, $B = 0$) and perfect conductors ($\sigma_{\text{wall}} \approx \infty$, $\partial B/\partial n = 0$) and for some problems a mixed type conductivity condition $\partial B/\partial n + cB = 0$, c being the conductivity constant parameter, will be imposed. Particular interest will be given to the case when one of the walls is partly insulating and partly perfectly conducting. In most of the problems, the no-slip velocity condition ($V = 0$) is imposed on the duct walls; however, the slip effect ($\alpha \partial V/\partial n + V = 0$, α being the slip length) is taken into consideration in some special cases of the problems.

The effects of the conducting portions on the walls, the increase in the value of Hartmann number, and the magnitude and behavior change caused by the slip (if it exists) on the flow and induced magnetic field will be discussed for each problem.

In most of the problems, the external magnetic field applies either horizontally (in the rectangular duct problems) or vertically (in the problems between parallel plates). For these cases, the MHD flow equations reduce to

$$\begin{aligned} \nabla^2 V + Ha\frac{\partial B}{\partial x} &= -1 + Re\frac{\partial V}{\partial t} \\ \nabla^2 B + Ha\frac{\partial V}{\partial x} &= R_m\frac{\partial B}{\partial t} \end{aligned} \tag{1.59}$$

and

$$\begin{aligned} \nabla^2 V + Ha\frac{\partial B}{\partial y} &= -1 + Re\frac{\partial V}{\partial t} \\ \nabla^2 B + Ha\frac{\partial V}{\partial y} &= R_m\frac{\partial B}{\partial t} \end{aligned}, \tag{1.60}$$

respectively, for horizontally and vertically applied magnetic fields.

When the steady MHD flow is considered for $Re << 1$ and $R_m << 1$, time derivative terms can be neglected, which is mostly the case in the pipe (channel) flow problems considered. Thus, the equations are simplified to [12]

$$\begin{aligned} \nabla^2 V + Ha\frac{\partial B}{\partial x} &= -1 \\ \nabla^2 B + Ha\frac{\partial V}{\partial x} &= 0 \end{aligned} \tag{1.61}$$

and

$$\begin{aligned} \nabla^2 V + Ha\frac{\partial B}{\partial y} &= -1 \\ \nabla^2 B + Ha\frac{\partial V}{\partial y} &= 0 \end{aligned} \tag{1.62}$$

again for horizontally and vertically applied magnetic fields, respectively.

When the flow is electrically driven in infinite channels by the probes attached to the infinite plates, the pressure gradient is taken as zero and the MHD flow equations become

$$\begin{aligned} \nabla^2 V + Ha\frac{\partial B}{\partial y} &= 0 \\ \nabla^2 B + Ha\frac{\partial V}{\partial y} &= 0 \end{aligned} \tag{1.63}$$

for which the applied magnetic field is perpendicular to the infinite plates.

The boundary element method (BEM) and its alternative form called the dual reciprocity boundary element method (DRBEM) that transforms the remaining domain integral to a boundary integral also are used for the numerical solution of the MHD channel flows. Since the boundary conductivity conditions are especially partly insulated and partly perfectly conducting type, the boundary element method fits best for handling these boundary conditions compared to other numerical methods. Perfectly conducting boundary conditions are Neumann type. In the discretized BEM systems, the unknowns contain both the solution and its normal derivative. Thus, the Neumann type boundary conditions are handled in BEM formulation by just inserting them to the final discretized system. BEM does not have to deal with a boundary term due to Neumann type boundary conditions as in the case of FEM. The BEM or DRBEM finds the unknown velocity and the induced magnetic field information (solution and its normal derivative) first on the discretized points of the boundary and then extends the solution to the arbitrarily selected interior points. It is not required to discretize the whole domain as in FEM. Thus, the significant changes in the flow and induced magnetic field behaviors due to the discontinuities on the boundary conditions are captured without difficulty by carefully arranging the discretization of the boundary of the problem region. These methods generate full coefficient matrices making them suitable only for models that are not too big.

The next chapter will present the BEM and DRBEM methods on the general Poisson type partial differential equation with Dirichlet and/or Neumann type boundary conditions. The derivation of the fundamental solution for coupled equations of the MHD duct steady flow is going to be given in details in Chap. 3 since it will be applied to all of the MHD flow problems considered in Chap. 4.

Chapter 4 will be denoted completely to the numerical results obtained by BEM for MHD duct flow problems governed by Eqs. (1.57) and its special forms given in Eqs. (1.59), (1.60), (1.61), (1.62), and (1.63) mostly for steady flows. The BEM applications will use the fundamental solution of coupled MHD equations. Problem domains are rectangular or circular ducts, infinite strips, half plane, or between parallel plates. Physically reasonable boundary conditions for the velocity V and the induced magnetic field B are used, and special emphasis is given on the partly insulated–partly perfectly conducting walls. Convergence of the BEM solutions in infinite regions and the computation of the parabolic boundary layer thickness emanating from the points where the conductivity on the same walls is changing are also presented.

Chapter 5 will contain the DRBEM solutions of the MHD flow problems together with the heat transfer. Mainly, the natural or mixed convection MHD flow, inductionless Buoyancy driven MHD flow, and inductionless MHD flow with electric potential, which are described in Sects. 1.2.4–1.2.7, are considered. The unidirectional time-dependent MHD duct flow problems given in Sect. 1.2.8 (Eq. (1.57)) are also solved with DRBEM in Sect. 5.1 since the fundamental solution required in BEM is not available.

Chapter 2
Boundary Element Solution of Potential Flow Problems

In this chapter, the numerical formulation of the boundary element method is presented first for the solution of potential flow problems, especially for the case of a function that is required to satisfy Poisson equation. In its more general form, this method consists of subdividing the boundary of the region under consideration into a series of elements. The boundary element method is a particular application of weighted residual techniques that is related to other numerical methods such as finite elements in that sense. The FEM requires a discretization of the entire domain, preserving the dimensional order of the problem. On the contrary, the BEM operates on the discretization of the boundary, which reduces the terms of the problem by one dimension (that is, for 2D problems only the line boundary of the domain needs to be discretized) [14–16]. Thus, the BEM is able to provide a complete problem solution in terms of boundary values (solution and its normal derivative) only, with considerable computer time and memory savings. The FEM results in very large system of equations, although sparse, for a fine mesh. However, the BEM gives much smaller system of equations to be solved, but full form with scattered zeros. The BEM requires the fundamental solution to the governing equations of the problem under consideration in order to avoid domain integrals in the formulation of the boundary integral equation. Only boundary integrals appear in the weighted residual statement. However, the nonhomogeneous and nonlinear terms are incorporated in the formulation by means of domain integrals. The use of cells to evaluate these domain integrals implies an internal discretization that considerably increases the quantity of data necessary to run a problem. Thus, the BEM loses the attraction of its boundary-only character in relation to the other domain decomposition methods. One of the several techniques to deal with the domain integrals is the dual reciprocity boundary element method (DRBEM), which is the subject of the second part of this chapter. It is essentially a generalized way of constructing particular solutions that can be used to solve nonlinear and time-dependent problems. The basic idea behind the DRBEM is to employ a fundamental solution corresponding to a simpler equation and to treat the remaining terms as

M. Tezer-Sezgin, C. Bozkaya, *Boundary Element Method for Magnetohydrodynamic Flow*, Surveys and Tutorials in the Applied Mathematical Sciences 14,
https://doi.org/10.1007/978-3-031-58353-7_2

inhomogeneity that is approximated by using radial basis functions. Then, the reciprocity principle is applied. Since only classical ideas of BEM and DRBEM are addressed in this chapter, the presentation is based on the well-known books of Brebbia [17], Brebbia and Dominguez [2], and Partridge et al. [3], Brebbia [18], Brebbia and Walker [19], Brebbia et al. [20], Katsikadelis [21], Pozrikidis [22], Paris and Canas [23], and Gibson [24]. Some other books on the boundary element method are also made use of as [25–30]. Beer et al. [31] focused on the programming aspects of BEM, and the linear algebra aspects of BEM are provided in Rjasanow and Steinbach [32].

2.1 BEM Formulation for Poisson Equation

The derivation of boundary element method goes back to 1978 with the publication of the proceedings of the first international conference on Boundary Elements [18]. Then, the fundamentals and some applications have been published in the pioneer books by the Computational Mechanics research group at Southampton [2, 17, 19, 20]. Pozrikidis [22] provides also BEMLIB software library as a practical guide to BEM.

This section starts with the derivation of the boundary integral equation of Poisson equation by using Green's second identity since the concentration will be on two-dimensional problems. The Poisson equation in three dimensions will be treated similarly by using divergence theorem, and the boundary integral equation is now over the surface of the volume region. Thus, the numerical solution process together with finding out the fundamental solution is discussed in some detail for two-dimensional problems using constant and linear elements. The fundamental solution in three dimensions and types of boundary elements on the surface of the volume are also going to be provided.

2.1.1 *Weighted Residuals for Poisson Equation*

The weighted residual technique is a general procedure in transforming differential equations into integral equations by minimizing the errors introduced in the problem from the replacement of the solution by its approximation.

Consider that we are seeking the solution u of the following Poisson equation in a two-dimensional or three-dimensional domain Ω [3, 17]

$$\nabla^2 u = f \tag{2.1}$$

with the partially given boundary conditions

$$\begin{aligned} u &= \bar{u} && \text{on } \Gamma_1 \\ q &= \frac{\partial u}{\partial n} = \bar{q} && \text{on } \Gamma_2, \end{aligned} \tag{2.2}$$

where n is the outward normal to the boundary $\Gamma = \partial\Omega$, with $\Gamma = \Gamma_1 \cup \Gamma_2$. The first condition in Eq. (2.2) is physically called *essential*, whereas the second is of *natural* type where $\bar{u}$ and $\bar{q}$ are known (given) values. More complex boundary conditions such as combinations of the two, i.e.,

$$au + bq = c,$$

can be easily included to the BEM formulation where a, b, and c are known parameters. The right-hand side function f in Eq. (2.1) may be a function of space and time, the unknown u itself or some of its space and time derivatives as nonlinear terms, or only a body force. In any case, it will be considered as the inhomogeneity in the Poisson equation (2.1). The function f will be assumed as a body force that is $f = f(x, y)$ or $f = f(x, y, z)$ in the BEM formulation, and the other cases of f are going to be treated by using the DRBEM approach in the last section of this chapter.

A weighting function u^* is introduced such that it is continuous up to the second derivatives within Ω, the normal derivative of which along the boundary is $q^* = \partial u^*/\partial n$. Replacing the solution u and its normal derivative q by their approximations, one can write the residuals obtained as

$$\begin{aligned} R &= \nabla^2 u \neq f && \text{in } \Omega \\ R_1 &= u - \bar{u} \neq 0 && \text{on } \Gamma_1 \\ R_2 &= q - \bar{q} \neq 0 && \text{on } \Gamma_2 \,. \end{aligned} \tag{2.3}$$

The above residuals can be weighted by the functions u^* and q^* (can be minimized by orthogonalization of them with respect to weight function and its normal derivative) giving the following weighted residual statement:

$$\int_\Omega (\nabla^2 u) u^* d\Omega = \int_{\Gamma_2} (q - \bar{q}) u^* d\Gamma - \int_{\Gamma_1} (u - \bar{u}) q^* d\Gamma + \int_\Omega f u^* \, d\Omega \,. \tag{2.4}$$

Application of Green's second identity two times in two dimensions or divergence theorem two times in three dimensions results in

$$\begin{aligned} \int_\Omega (\nabla^2 u^*) u d\Omega = &- \int_{\Gamma_2} \bar{q} u^* d\Gamma - \int_{\Gamma_1} q u^* d\Gamma + \int_{\Gamma_2} u q^* d\Gamma \\ &+ \int_{\Gamma_1} \bar{u} q^* d\Gamma + \int_\Omega f u^* \, d\Omega \,. \end{aligned} \tag{2.5}$$

Now, we want to transform formula (2.5) into a boundary integral equation except the domain or volume integral containing the inhomogeneity f. Its transformation to boundary integral will be shown in Sect. 2.2 using the dual reciprocity principle. The transformation of the integral over Ω on the left-hand side of Eq. (2.5) into a boundary integral is done by a special type of weighting function u^* called the fundamental solution. That is, find a solution satisfying the Laplace equation at a source point i whose effect is propagated to infinity without any consideration of boundary conditions. Thus, u^* governs the equation [2]

$$\nabla^2 u^* + \Delta_i = 0, \tag{2.6}$$

where Δ_i represents Dirac delta function that goes to infinity at the point i and is equal to zero elsewhere. With the property of Dirac delta function

$$\int_\Omega u(\nabla^2 u^*)d\Omega = \int_\Omega u(-\Delta_i)d\Omega = -u_i. \tag{2.7}$$

Equation (2.5) can be written now as

$$u_i + \int_{\Gamma_2} uq^* d\Gamma + \int_{\Gamma_1} \bar{u}q^* d\Gamma = \int_{\Gamma_2} \bar{q}u^* d\Gamma + \int_{\Gamma_1} qu^* d\Gamma - \int_\Omega u^* f d\Omega. \tag{2.8}$$

In Eq. (2.8), the values of u^* and q^* correspond only to the position of the source point i. For each different position, a new integral equation is written.

For an isotropic two-dimensional or three-dimensional medium, the fundamental solutions of Eq. (2.6) are [2]

$$u^* = \frac{1}{2\pi}\ln\left(\frac{1}{r}\right) \text{ or } u^* = \frac{1}{4\pi r}, \tag{2.9}$$

where r is the distance from the source point i to any other point in the region. The Laplace equation in polar or spherical coordinates assuming axial symmetry is satisfied by the fundamental solution u^* as

$$\nabla^2 u^* = \frac{\partial^2 u^*}{\partial r^2} + \frac{\alpha}{r}\frac{\partial u^*}{\partial r} = -\Delta_i \tag{2.10}$$

for any value of $r \neq 0$, and $\alpha = 1$ in 2D, $\alpha = 2$ in 3D. For the case of $r = 0$, one can integrate around a small circle in 2D or small sphere in 3D, denoted by Γ_ϵ with radius ϵ, and then take the limit when $\epsilon \to 0$, i.e.,

$$\int_{\Omega_\epsilon} (\nabla^2 u^*)d\Omega = \int_{\Gamma_\epsilon} \frac{\partial u^*}{\partial n} d\Gamma = \int_{\Gamma_\epsilon} \frac{\partial u^*}{\partial r} d\Gamma \tag{2.11}$$

from the divergence theorem. Now,

$$\lim_{\epsilon\to 0}\int_{\Gamma_\epsilon}\frac{\partial u^*}{\partial r}d\Gamma = -\lim_{\epsilon\to 0}\int_{\Gamma_\epsilon}\frac{1}{2\pi\epsilon}d\Gamma = -\lim_{\epsilon\to 0}\frac{2\pi\epsilon}{2\pi\epsilon} = -1 \text{ in 2D} \tag{2.12}$$

or

$$\lim_{\epsilon\to 0}\int_{\Gamma_\epsilon}\frac{\partial u^*}{\partial r}d\Gamma = -\lim_{\epsilon\to 0}\int_{\Gamma_\epsilon}\frac{1}{4\pi\epsilon^2}d\Gamma = -\lim_{\epsilon\to 0}\frac{4\pi\epsilon^2}{4\pi\epsilon^2} = -1 \text{ in 3D} \tag{2.13}$$

with the perimeter of the small circle $\Gamma_\epsilon = 2\pi\epsilon$ or the surface area of the small sphere $\Gamma_\epsilon = 4\pi\epsilon^2$. Thus,

$$\int_\Omega(\nabla^2 u^*)d\Omega = \int_\Omega(-\Delta_i)d\Omega = -1\,.$$

Equation (2.8) is valid for any source point in the domain Ω. Since the aim is to solve equation (2.8) first for the boundary points, it is necessary to see the form of the equation when the point i is on Γ. For this, we consider now a semicircle $\Gamma_\epsilon^{'}$ of radius ϵ in 2D or a hemisphere $\Gamma_\epsilon^{'}$ of radius ϵ in 3D around the center point i and then reduce ϵ to zero in the limit sense. The boundary is assumed to be smooth and the point i to be on Γ_2 (similar considerations apply if it is on Γ_1). Considering Eq. (2.8) without applying the boundary conditions, we may write

$$u_i + \int_\Gamma uq^*\,d\Gamma + \int_\Omega fu^*\,d\Omega = \int_\Gamma u^*q\,d\Gamma\,. \tag{2.14}$$

Now, the integrals over Γ (Γ_2 or Γ_1) are divided into two pieces as

$$\int_{\Gamma-2\epsilon} u\frac{\partial u^*}{\partial n}\,d\Gamma + \int_{\Gamma_\epsilon^{'}} u\frac{\partial u^*}{\partial n}\,d\Gamma \tag{2.15}$$

and

$$\int_{\Gamma-2\epsilon} u^*\frac{\partial u}{\partial n}\,d\Gamma + \int_{\Gamma_\epsilon^{'}} u^*\frac{\partial u}{\partial n}\,d\Gamma \tag{2.16}$$

with the perimeter $\Gamma_\epsilon^{'} = \pi\epsilon$ of the semicircle in 2D, and the half of the hemisphere $\Gamma_\epsilon^{'} = 2\pi\epsilon^2$ in 3D, respectively.

When ϵ tends to zero, the limiting process in 2D

$$\lim_{\epsilon\to 0}\int_{\Gamma_\epsilon^{'}}\frac{\partial u}{\partial n}u^*\,d\Gamma = \lim_{\epsilon\to 0}\int_{\Gamma_\epsilon^{'}}\frac{-\ln\epsilon}{2\pi}q\,d\Gamma = \lim_{\epsilon\to 0}\left\{\frac{-\pi\epsilon\ln\epsilon}{2\pi}q\right\} = 0 \tag{2.17}$$

$$\lim_{\epsilon\to 0}\int_{\Gamma_\epsilon^{'}} u\frac{\partial u^*}{\partial n}\,d\Gamma = \lim_{\epsilon\to 0}\int_{\Gamma_\epsilon^{'}}(-\frac{1}{2\pi\epsilon})u\,d\Gamma = \lim_{\epsilon\to 0}(-u\frac{\pi\epsilon}{2\pi\epsilon}) = -\frac{1}{2}u_i \tag{2.18}$$

or in 3D

$$\lim_{\epsilon\to 0}\int_{\Gamma'_\epsilon}\frac{\partial u}{\partial n}u^*\,d\Gamma=\lim_{\epsilon\to 0}\int_{\Gamma'_\epsilon}\frac{1}{4\pi\epsilon}q\,d\Gamma=\lim_{\epsilon\to 0}\left\{\frac{-2\pi\epsilon^2}{4\pi\epsilon}q\right\}=0 \tag{2.19}$$

$$\lim_{\epsilon\to 0}\int_{\Gamma'_\epsilon}u\frac{\partial u^*}{\partial n}\,d\Gamma=\lim_{\epsilon\to 0}\int_{\Gamma'_\epsilon}(-\frac{1}{4\pi\epsilon^2})u\,d\Gamma=\lim_{\epsilon\to 0}(-u\frac{2\pi\epsilon^2}{4\pi\epsilon^2})=-\frac{1}{2}u_i \tag{2.20}$$

shows the continuity of integrals in Eqs. (2.8) and (2.14) when the point i is on the boundary.

Substituting now the results (2.15)–(2.18) in Eq. (2.8) for two dimensions or (2.15)–(2.16) and (2.19)–(2.20) in Eq. (2.8) for three dimensions for a boundary point i, the following boundary integral equation gives the starting point of the boundary element method as

$$\frac{1}{2}u_i+\int_{\Gamma_2}u\frac{\partial u^*}{\partial n}\,d\Gamma+\int_{\Gamma_1}\bar{u}\frac{\partial u^*}{\partial n}\,d\Gamma=\int_{\Gamma_2}\bar{q}u^*\,d\Gamma+\int_{\Gamma_1}\frac{\partial u}{\partial n}u^*\,d\Gamma-\int_{\Omega}fu^*\,d\Omega \tag{2.21}$$

in which the unknowns u on Γ_2 and $\frac{\partial u}{\partial n}$ on Γ_1 can be solved since $u=\bar{u}$ on Γ_1 and $\frac{\partial u}{\partial n}=q=\bar{q}$ on Γ_2 are given.

2.1.2 Boundary Elements

Equation (2.21) is going to be discretized to find a system of equations from which the unknown boundary values can be found when Ω is a two-dimensional region. The boundary Γ of the domain Ω is divided into N straight-line segments or elements.

The points (nodes), where the unknown values are considered, are taken at the midpoints of the line segments to give constant elements and at the intersections between two elements for giving linear elements as shown in Fig. 2.1.

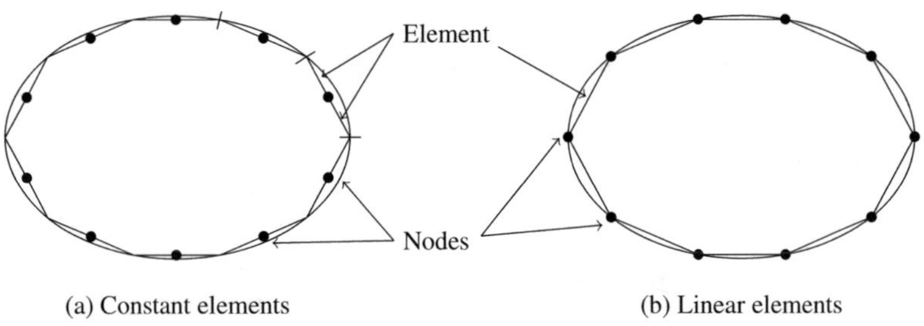

Fig. 2.1 Constant and linear boundary elements in two dimensions

For the constant element case, Eq. (2.21) can be discretized for a given point i before applying any boundary conditions, as

$$\frac{1}{2}u_i + \sum_{j=1}^{N} u_j \int_{\Gamma_j} q^* \, d\Gamma + \int_{\Omega} f u^* \, d\Omega = \sum_{j=1}^{N} q_j \int_{\Gamma_j} u^* \, d\Gamma \tag{2.22}$$

since the values of u and q are assumed to be constant over each element and equal to the values at the midpoint of the element.

The two types of integrals in Eq. (2.22) that are $\int_{\Gamma_j} q^* \, d\Gamma$ and $\int_{\Gamma_j} u^* \, d\Gamma$ relate the boundary node i where the fundamental solution is applied to any other node j. Thus, their values are denoted as

$$\bar{H}_{ij} = \int_{\Gamma_j} q^* \, d\Gamma, \quad G_{ij} = \int_{\Gamma_j} u^* \, d\Gamma \tag{2.23}$$

giving the discretized form of Eq. (2.21)

$$\frac{1}{2}u_i + \sum_{j=1}^{N} \bar{H}_{ij} u_j + C_i = \sum_{j=1}^{N} G_{ij} q_j, \quad i = 1, \ldots, N . \tag{2.24}$$

The value $1/2$ is summed to the diagonal entries of the matrix $\bar{H}$ by writing

$$H_{ij} = \bar{H}_{ij} + \frac{1}{2}\delta_{ij}, \tag{2.25}$$

where δ is the Kronecker delta function, and Eq. (2.24) can now be written as

$$C_i + \sum_{j=1}^{N} H_{ij} u_j = \sum_{j=1}^{N} G_{ij} q_j, \quad i = 1, \ldots, N , \tag{2.26}$$

where the term C_i is the result of the numerical integration of the domain integral and is different for each source point i on the boundary.

A system of equations is obtained resulting from the application of (2.26) to each boundary node that can be expressed in matrix–vector form as (assembly for N boundary elements)

$$\mathbf{Hu} = \mathbf{Gq} - \mathbf{C}, \tag{2.27}$$

where $\mathbf{H}$ and $\mathbf{G}$ are two $N \times N$ matrices and $\mathbf{u}$ and $\mathbf{q}$ are vectors of size $N \times 1$. Since N_1 values of $\mathbf{u}$ and N_2 values of $\mathbf{q}$ are known on Γ_1 and Γ_2, respectively ($N_1 + N_2 = N$), there are only N unknowns in the system of Eq. (2.27). To impose the boundary conditions (2.2) into the system (2.27), one has to rearrange the system by moving columns of $\mathbf{H}$ and $\mathbf{G}$ from one side to the other. Once all unknowns are passed to the left-hand side, one can write a system of equations

$$\mathbf{Ax} = \mathbf{y}, \tag{2.28}$$

where $\mathbf{x}$ is a vector containing unknown values of $\mathbf{u}$ and $\mathbf{q}$, and $\mathbf{y}$ is found by multiplying the corresponding columns of $\mathbf{H}$ and $\mathbf{G}$ by the known values of $\mathbf{u}$ and $\mathbf{q}$, and adding to the entries of the vector $\mathbf{C}$.

The solution of the system (2.28) gives unknown values of $\mathbf{u}$ and $\mathbf{q}$ on the boundary nodes, that is, all the boundary values are now known. Then, it is possible to calculate internal values of $\mathbf{u}$ or its derivatives. The values of $\mathbf{u}$ are calculated at any internal point i using Eq. (2.8) as

$$u_i = \int_\Gamma q u^* \, d\Gamma - \int_\Gamma u q^* \, d\Gamma - \int_\Omega f u^* \, d\Omega, \quad i = 1, \ldots, L \tag{2.29}$$

and with the same discretization for the boundary integrals, one arrives at L interior points

$$u_i = \sum_{j=1}^{N} G_{ij} q_j - \sum_{j-1}^{N} \bar{H}_{ij} u_j - C_i, \quad i = 1, \ldots, L\,. \tag{2.30}$$

In the evaluation of integrals in $\bar{H}_{ij}$ and G_{ij} for Eq. (2.30), the distance r is now measured from the interior point i to the boundary point j.

The space derivatives at an interior point i can also be calculated by differentiating equation (2.29)

$$\frac{\partial u}{\partial x}|_i = \int_\Gamma q \frac{\partial u^*}{\partial x} \, d\Gamma - \int_\Gamma u \frac{\partial q^*}{\partial x} \, d\Gamma - \int_\Omega \frac{\partial f}{\partial x} u^* \, d\Omega - \int_\Omega f \frac{\partial u^*}{\partial x} \, d\Omega,$$
$$i = 1, \ldots, L \tag{2.31}$$

$$\frac{\partial u}{\partial y}|_i = \int_\Gamma q \frac{\partial u^*}{\partial y} \, d\Gamma - \int_\Gamma u \frac{\partial q^*}{\partial y} \, d\Gamma - \int_\Omega \frac{\partial f}{\partial y} u^* \, d\Omega - \int_\Omega f \frac{\partial u^*}{\partial y} \, d\Omega,$$
$$i = 1, \ldots, L \tag{2.32}$$

for a constant element approximation over Γ.

2.1.3 *Integral Computations*

The evaluation of integrals in the entries $\bar{H}_{ij}$ and G_{ij} given in (2.23) is performed in two dimensions using numerical integration like Gaussian quadrature for the case $i \neq j$. When $i = j$, the singularity of the fundamental solution as $r \to 0$ requires a special formula taking care of the singularity (i.e., logarithmic or other transformations). For the constant element case, the diagonal entries H_{ii} and G_{ii}

can be computed analytically. $\bar{H}_{ii}$ are identically zero since the distance vector $\mathbf{r}$ and the normal $\mathbf{n}$ are perpendicular to each other, giving $H_{ii} = 1$. The diagonal entries of the matrix $\mathbf{G}$ are

$$G_{ii} = \int_{\Gamma_i} u^* \, d\Gamma = \frac{1}{2\pi} \int_{\Gamma_i} \ln(\frac{1}{r}) \, d\Gamma, \quad i = 1, \ldots, N \, . \tag{2.33}$$

With the change of variables $r = |\frac{l}{2}\xi|$, $d\Gamma = dr = \frac{l}{2} d\xi$, where l is the length of the constant element, and G_{ii} becomes

$$G_{ii} = \frac{1}{\pi} \int_{node\, i}^{end\ point} \ln(\frac{1}{r}) \, dr = \frac{l}{2\pi} \int_0^1 \ln(\frac{2}{l\xi}) \, d\xi = \frac{l}{2\pi} (\ln(\frac{2}{l}) + 1) \tag{2.34}$$

by using the symmetry from node i to the end points of the element.

When the right-hand side function f is a known function of space variables as $f = f(x, y)$ in the Poisson equation (2.1), it results in a domain integral in the weighted residual statement (2.4), the computed value C_i in the element discretized equation (2.26) when i is a boundary point, and finally a vector $\mathbf{C}$ in the boundary element discretized system (2.27) with the entries C_i. The integration over the domain Ω can be performed by dividing Ω into a series of cells, and over each cell a numerical integration formula, i.e., Gauss quadrature, can be applied. Thus, the entries of the vector $\mathbf{C}$ are computed as

$$C_i = \sum_{e=1}^{E} (\sum_{k=1}^{K} w_k (f u^*)_k) \Omega_e , \tag{2.35}$$

where E is the number of cells in Ω, w_k are the Gaussian quadrature weights, fu^* is going to be evaluated at integration points k, k varying up to the number of integration points K, and Ω_e is the area of the cell.

If the interior region is divided into rectangles with equally spaced points in x- and y-directions, composite Trapezoidal rule can also be used forming larger cells with the same discretization.

The domain integrals need to be computed also when interior solution is required in Eqs. (2.29) and (2.30). In these integral evaluations, the point i is in Ω, and fu^* is evaluated again at the integration point k in the numerical integration (2.35).

Since the computation of the domain integral destroys the main advantage of the boundary element method causing significant computational load, particular solution is sought for transforming the Poisson equation into the Laplace equation. When the particular solution cannot be obtained easily, a series of approximate particular solutions can be found by using a series of radial basis functions. These are the procedures in the dual reciprocity boundary element method that will be discussed in Sect. 2.2.

2.1.3.1 Constant and Linear Element Interpolations

Boundary elements will be given first for the boundary Γ of a two-dimensional domain Ω. The boundary integral equation (2.21) is written as

$$\frac{1}{2}u_i + \int_\Gamma u\frac{\partial u^*}{\partial n}\,d\Gamma = \int_\Gamma \frac{\partial u}{\partial n}u^*\,d\Gamma - \int_\Omega fu^*\,d\Gamma \tag{2.36}$$

when the point $i(=1,\ldots,N)$ is on the boundary Γ and it is reduced to the computation of the solution as in Eq. (2.29)

$$u_i = \int_\Gamma \frac{\partial u}{\partial n}u^*\,d\Gamma - \int_\Gamma u\frac{\partial u^*}{\partial n}\,d\Gamma - \int_\Omega fu^*\,d\Gamma \tag{2.37}$$

at an interior point $i(=1,\ldots,L)$.

Now, Eqs. (2.36)–(2.37) can be represented using the coefficient of the solution u as c_i [2]

$$c_iu_i + \int_\Gamma u\frac{\partial u^*}{\partial n}\,d\Gamma = \int_\Gamma \frac{\partial u}{\partial n}u^*\,d\Gamma - \int_\Omega fu^*\,d\Omega, \tag{2.38}$$

where

$$c_i = \frac{\theta}{2\pi} \tag{2.39}$$

θ being the internal angle at the point i. The constant $c_i = 1/2$ applies only for a smooth boundary, while $c_i = \theta/2\pi$ can be obtained by defining a small spherical or circular region around the corners and taking the radius of them to zero. The evaluation of the singular integral in Eq. (2.18) gives

$$\lim_{\epsilon\to 0}\int_{\Gamma_\epsilon} u\frac{\partial u^*}{\partial n}\,d\Gamma = -u_i \tag{2.40}$$

for an interior point i since the ϵ-sector is a circle with perimeter $2\pi\epsilon$. Thus, the formula (2.38) represents both cases of the position (geometry) of the point i.

In the boundary integral equation (2.38), the solution u and its normal derivative $q = \partial u/\partial n$ that are the unknowns on Γ_2 and Γ_1, respectively, need interpolatory forms for each boundary element using either constant or linear approximations. For this numerical purpose, a point placed somewhere along the element should be expressed in terms of local coordinates along the element taking values ± 1 at the element edges and zero at the middle of the element [17].

Thus, for a constant approximation of u and q over an element, we can write with a change of variable $\xi = \frac{x}{l/2}$ over $(-\frac{l}{2} \le x \le \frac{l}{2})$

$$u = N_1(\xi)u_1, \quad q = N_1(\xi)q_1, \tag{2.41}$$

where $N_1(\xi) = 1$ over the constant element and u_1, q_1 are the values at the middle point and also assumed to be the same over the element. For a linear approximation of u and q over an element

$$\begin{aligned} u &= N_1(\xi)u_1 + N_2(\xi)u_2 \\ q &= N_1(\xi)q_1 + N_2(\xi)q_2 \end{aligned}, \tag{2.42}$$

where

$$N_1(\xi) = \frac{1-\xi}{2}, \quad N_2 = \frac{1+\xi}{2}, \quad -1 \leq \xi \leq 1, \tag{2.43}$$

and u_1, q_1, and u_2, q_2 are the values of u and $\partial u/\partial n$ at the end points 1 and 2, respectively, of the linear element. The constant function $N_1(\xi)$ or linear functions $N_1(\xi)$, $N_2(\xi)$ are called the shape functions in the constant and linear interpolations (2.41) and (2.42), respectively. Higher-order elements can be generated similarly.

In the constant element discretization of the boundary, the u_j and q_j $(j = 1, \ldots, N)$ can be taken out of the integrals Γ_j as is done in Eq. (2.22), and only $\int_{\Gamma_j} u^*\, d\Gamma$ and $\int_{\Gamma_j} q^*\, d\Gamma$ are computed. These integrals give the entries of $\bar{H}_{ij}$, G_{ij}, and H_{ij} as given in Eqs. (2.23) and (2.25), respectively, and the summation for each boundary element (assembly) results in the system $\mathbf{Hu} = \mathbf{Gq} - \mathbf{C}$ as in Eq. (2.27).

However, when the boundary discretization is performed with linear elements and interpolations of u and q are taken as in Eq. (2.42), the variations of u and q are linear, and they cannot be taken out of the integrals in Eq. (2.38) since

$$c_i u_i + \sum_{j=1}^{N} \int_{\Gamma_j} u q^*\, d\Gamma = \sum_{j=1}^{N} \int_{\Gamma_j} q u^*\, d\Gamma - \int_{\Omega} f u^*\, d\Omega, \quad i = 1, \ldots, N, \tag{2.44}$$

and for each integral over Γ_j, we have two components in $\bar{H}_{ij}$ and G_{ij}

$$h_{i1} = \int_{\Gamma_j} N_1 q^*\, d\Gamma,\ h_{i2} = \int_{\Gamma_j} N_2 q^*\, d\Gamma,\ g_{i1} = \int_{\Gamma_j} N_1 u^*\, d\Gamma,$$

$$g_{i2} = \int_{\Gamma_j} N_2 u^*\, d\Gamma\,. \tag{2.45}$$

Thus, the entries of the matrix $\bar{H}_{ij}$ are the sum of the terms h_{i1} coming from the jth element and h_{i2} coming from the $(j-1)$th element. Similarly, the G_{ij} matrix is formed. Using again Eq. (2.25) $H_{ij} = \bar{H}_{ij} + \frac{1}{2}\delta_{ij}$, the boundary element discretized form of Eq. (2.44) becomes

$$\sum_{j=1}^{N} H_{ij} u_j + C_i = \sum_{j=1}^{N} G_{ij} q_j, \quad i = 1, \ldots, N, \tag{2.46}$$

and the whole system of matrix–vector equations

$$\mathbf{Hu} = \mathbf{Gq} - \mathbf{C} \tag{2.47}$$

is obtained for a linear element approximation of u and q.

When the boundary Γ is not smooth, the value $c_i = 1/2$ is not valid. For this case, the diagonal entries of the matrix $\mathbf{H}$ can be computed as

$$h_{ii} = -\sum_{i=1, i\neq j}^{N} h_{ij} \tag{2.48}$$

with the assumption that a uniform potential is applied over the whole boundary giving the normal derivatives (q values) as zero which is $\mathbf{Hu} = \mathbf{0}$.

The problem of having two unknown fluxes at a corner node, node 2 of element "j" and node 1 of element "$j+1$," can be shifted inside the two linear elements. The solution and its flux are represented by linear functions along the whole elements in terms of their nodal values, and both of them are discontinuous at the corner. The total number of nodes will be equal to the total number of elements plus one additional node per each discontinuous element [2].

For arbitrary geometries, some types of curvilinear elements (e.g., three-noded quadratic elements, four-noded cubic elements, or two-noded cubic elements with u and q variations at the two end points of the element) can be used [2]. Boundary elements used in three-dimensional problems are surface elements that cover the boundary of the body. These are either triangular or quadrilateral elements that can be flat or curved. The functions u and q can be constant, or varying linearly over the element, or being second-order functions that produce a curved element [2]. Rjasanow and Steinbach [32] have been provided piecewise constant, piecewise linear discontinuous, and continuous basis functions in 3D.

2.2 The Dual Reciprocity Boundary Element Method for Poisson Equation of the Type $\nabla^2 u = b(x, y, u, u_x, u_y, \partial u/\partial t)$

The BEM application of a partial differential equation requires a fundamental solution that takes into account all the terms in the governing equation. Usually such a fundamental solution is difficult to obtain, and a fundamental solution to a simpler form of the differential equation is used. This results in a domain integral in the corresponding integral equation, and the domain integral has to be calculated using internal cells destroying the boundary-only nature of the BEM.

The dual reciprocity boundary element method was first proposed in [33]. It may be used with a fundamental solution of a simpler differential operator in the equation and does not need internal cells; however, an arbitrary number of interior nodes are used for computing the solution inside the region Ω.

2.2.1 The Dual Reciprocity Principle for the Poisson Equation $\nabla^2 u = b$

The dual reciprocity method idea is going to be explained for the Poisson equation as given in [3]

$$\nabla^2 u = b \tag{2.49}$$

in which the right-hand side function may be a known function $b = b(x, y)$ or a nonlinear function of u, $b = b(x, y, u, u_x, u_y)$, or a time-dependent function $b = b(x, y, t, u, u_t)$.

If a particular solution $\hat{u}$ is available for Poisson equation (2.49), then $\nabla^2 \hat{u} = b$ and $\nabla^2 u^h = 0$, where $u = u^h + \hat{u}$. It is generally difficult to find particular solution for nonlinear and time-dependent problems. The dual reciprocity method uses a series of particular solutions $\hat{u}_j$ instead of a single particular solution $\hat{u}$ [3].

First, the right-hand side function b is approximated by using radial basis functions f_j as

$$b = \sum_{j=1}^{N+L} \alpha_j f_j, \tag{2.50}$$

where α_j are a set of initially unknown coefficients, and N and L are the number of boundary and interior nodes, respectively. The radial basis function f_j is linked to particular solution $\hat{u}_j$ through

$$\nabla^2 \hat{u}_j = f_j \,. \tag{2.51}$$

The functions f_j are geometry dependent and may be a polynomial, exponential, or logarithmic function in the distance r used in the fundamental solution. The distance vector $\mathbf{r}$ is written with its components $\mathbf{r} = (r_x, r_y)$ in x- and y-directions. The distance r is given as

$$r^2 = r_x^2 + r_y^2, \tag{2.52}$$

and the radial basis function f is usually taken as a polynomial in r [3], that is,

$$f = f(r) = 1 + r + r^2 + r^3 + \cdots + r^m \,, \tag{2.53}$$

where m is the degree of the polynomial $f(r)$.

The corresponding particular solution $\hat{u}$ and its normal derivative $\hat{q}$ are obtained as

$$\hat{u} = \frac{r^2}{4} + \frac{r^3}{9} + \cdots + \frac{r^{m+2}}{(m+2)^2} \tag{2.54}$$

with axisymmetric Laplace operator $\nabla^2 = \frac{\partial^2}{\partial r^2} + \frac{1}{r}\frac{\partial}{\partial r}$. Then,

$$\hat{q} = \frac{\partial \hat{u}}{\partial r}\frac{\partial r}{\partial x}\frac{\partial x}{\partial n} + \frac{\partial \hat{u}}{\partial r}\frac{\partial r}{\partial y}\frac{\partial y}{\partial n} = (r_x \frac{\partial x}{\partial n} + r_y \frac{\partial y}{\partial n})(\frac{1}{2} + \frac{r}{3} + \cdots + \frac{r^m}{(m+2)}), \tag{2.55}$$

where $\partial r/\partial x = r_x/r$, $\partial r/\partial y = r_y/r$, and $\partial x/\partial n = n_x = \cos(\mathbf{n}, x)$, $\partial y/\partial n = n_y = \cos(\mathbf{n}, y)$ are the direction cosines with respect to x- and y-axes with the outward normal $\mathbf{n} = (n_x, n_y)$ at the boundary.

On the other hand, different kinds of radial basis functions can be also used for the Laplace operator that results in corresponding particular solution $\hat{u}$ as follows [3, 34, 35]:

$$f = r^2 \ln r, \qquad \hat{u} = \frac{r^4}{16}(\ln r - \frac{1}{2})$$
$$f = e^{-r^2}, \qquad \hat{u} = \frac{1}{4}(\ln r^2 + E_1(r^2)) \text{ with } E_1(x) = \int_x^\infty \frac{e^{-t}}{t}\,dt$$
$$f = \frac{2c - r}{(r+c)^4}, \quad \hat{u} = -\frac{c + 2r}{2(r+c)^2},$$

where c is an arbitrary constant.

Substitution of Eq. (2.51) into Eq. (2.50) and then into original Poisson equation (2.49) replaces the right-hand side function b with an approximation in terms of Laplacian of particular solutions as follows:

$$\nabla^2 u = \sum_{j=1}^{N+L} \alpha_j (\nabla^2 \hat{u}_j) = b\,. \tag{2.56}$$

The BEM procedure with the fundamental solution of Laplace equation will now be applied to both sides of Eq. (2.56). That is, Eq. (2.56) is multiplied by the fundamental solution given in Eq. (2.9) and integrated over the domain, giving

$$\int_\Omega (\nabla^2 u) u^*\, d\Omega = \sum_{j=1}^{N+L} \alpha_j \int_\Omega (\nabla^2 \hat{u}_j) u^*\, d\Omega\,. \tag{2.57}$$

Application of Green's second identity to both sides produces the boundary integral equation for a source point i ($i = 1, \ldots, N$)

$$c_i u_i + \int_\Gamma q^* u \, d\Gamma - \int_\Gamma u^* q \, d\Gamma = \sum_{j=1}^{N+L} \alpha_j (c_i \hat{u}_{ij} + \int_\Gamma q^* \hat{u}_j \, d\Gamma - \int_\Gamma u^* \hat{q}_j \, d\Gamma), \tag{2.58}$$

where $\hat{q}_j = \dfrac{\partial \hat{u}_j}{\partial n} = \dfrac{\partial \hat{u}_j}{\partial x}\dfrac{\partial x}{\partial n} + \dfrac{\partial \hat{u}_j}{\partial y}\dfrac{\partial y}{\partial n}$.

Thus, the right-hand side function b in Eq. (2.49) has contributed to the BEM equation with an equivalent boundary integral instead of domain integral. This has been managed by introducing Laplace operator to the approximation of the function b and applying reciprocity principle to both sides of Eq. (2.56); hence, the dual reciprocity boundary element methods are used.

After discretizing the boundary Γ with either constant or linear elements given in (2.41) or (2.42), respectively, and evaluating the boundary integrals over each element, Eq. (2.58) can be written for all N boundary and arbitrarily selected L interior nodes as

$$c_i u_i + \sum_{k=1}^{N} H_{ik} u_k - \sum_{k=1}^{N} G_{ik} q_k = \sum_{j=1}^{N+L} \alpha_j (c_i \hat{u}_{ij} + \sum_{k=1}^{N} H_{ik} \hat{u}_{kj} - \sum_{k=1}^{N} G_{ik} \hat{q}_{kj}), \tag{2.59}$$

for $i = 1, \ldots, N$. Applying assembly procedure to Eq. (2.59) for N boundary elements, the following BEM discretized matrix–vector equation is obtained

$$\mathbf{Hu} - \mathbf{Gq} = \sum_{j=1}^{N+L} (\mathbf{H}\hat{\mathbf{u}}_j - \mathbf{G}\hat{\mathbf{q}}_j), \tag{2.60}$$

where $\mathbf{H}$ and $\mathbf{G}$ matrices are the same $N \times N$ matrices shown in Eq. (2.27) and defined by Eqs. (2.23) and (2.25). The constant $c_i = \theta/2\pi$ is also the same as given in (2.39) with an internal angle θ. The interior nodes where $c_i = 1$ may be located in any order at the L points where the interior solution is desired. For a straight boundary node i, $c_i = \frac{1}{2}$ is taken in Eq. (2.59).

Now, the matrices $\hat{\mathbf{U}}$ and $\hat{\mathbf{Q}}$ are formed from the vectors $\hat{\mathbf{u}}_j$ and $\hat{\mathbf{q}}_j$ placed as columns, and thus, they are of size $N \times (N + L)$. A vector $\boldsymbol{\alpha}$ will also be formed from the coefficients α_j in the approximation

$$\mathbf{b} = \sum_{j=1}^{N+L} \alpha_j f_j, \tag{2.61}$$

which is the solution of the $(N + L) \times (N + L)$ system

$$\mathbf{b} = \mathbf{F}\boldsymbol{\alpha}, \tag{2.62}$$

where the vector $\mathbf{b}$ involves the values of b, i.e., $b = b(x, y)$ at $N + L$ points, and each column of the matrix $\mathbf{F}$ consists of a vector $\mathbf{f}_j$ evaluated at $N + L$ points with the function f_j and $\boldsymbol{\alpha}$ is the $(N+L) \times 1$ vector with entries α_j. With all these newly defined $\hat{\mathbf{U}}$, $\hat{\mathbf{Q}}$, $\mathbf{F}$ matrices and $\boldsymbol{\alpha}$, $\mathbf{b}$ vectors, Eq. (2.60) is expressed as a matrix–vector equation of size $N \times 1$

$$\mathbf{Hu} - \mathbf{Gq} = (\mathbf{H}\hat{\mathbf{U}} - \mathbf{G}\hat{\mathbf{Q}})\boldsymbol{\alpha} \tag{2.63}$$

and

$$\mathbf{Hu} - \mathbf{Gq} = (\mathbf{H}\hat{\mathbf{U}} - \mathbf{G}\hat{\mathbf{Q}})\mathbf{F}^{-1}\mathbf{b}\,. \tag{2.64}$$

The matrix $\mathbf{F}$ is a coordinate matrix constructed column-wise with the vectors $\mathbf{f}_j$ having the function values $f_{ij} = 1 + r_{ij} + \cdots + r_{ij}^m$ with a distance r_{ij} between the points i and j $(i, j = 1, \ldots, N + L)$. Thus, $\mathbf{F}$ is a $(N + L) \times (N + L)$ symmetric and nonsingular matrix for which the system (2.62) is solvable. Matrix $\mathbf{F}$ depends only on geometric data and has no relation to either governing equation or boundary conditions. It may be calculated once and stored involving the same discretization. The calculation of $\mathbf{F}^{-1}$ poses no problems if the constant is included in the $f(r)$ expansion (2.53) and if no two nodes coincide. The simplest $f(r)$ expansion, $f(r) = 1+r$, produces quite sufficient results for Poisson type equations with $b(x, y, t, u, u_x, u_y, u_t)$. It is unnecessary to increase the order of the $f(r)$ expansion for this case. For the nonlinear terms in b, the order of $f(r)$ expansion may be increased to two or three but making no significant improvement in the results.

Applying boundary conditions (N_1 known entries of $\mathbf{u}$ and N_2 known entries of $\mathbf{q}$ with the assumption that Γ_1 and Γ_2 contain N_1 and N_2 nodes, respectively, in boundary conditions (2.2)) to Eq. (2.64), it can be reduced to the $N \times 1$ system

$$\mathbf{Ax} = \mathbf{b}\,. \tag{2.65}$$

The solution $\mathbf{x}$ of Eq. (2.65) gives unknown solution values of u and unknown normal derivative values q on the boundary, that is, the boundary solution is complete. The values of the solution u at any interior point $i (= 1, \ldots, L)$ are obtained from Eq. (2.59) by taking $c_i = 1$.

Alternatively, Eq. (2.59) can be written for all boundary $(c_i = 1/2)$ and interior $(c_i = 1)$ nodes by taking $i = 1, \ldots, N + L$, and the sizes of the matrices in the system (2.64) are matched. To do this, the matrices $\mathbf{H}$, $\mathbf{G}$ and $\hat{\mathbf{U}}$, $\hat{\mathbf{Q}}$ are enlarged to have sizes $(N+L)\times(N+L)$ by filling matrix partitions with the ones corresponding to the boundary and interior solutions in matrices $\mathbf{H}$, $\mathbf{G}$, $\hat{\mathbf{U}}$ and $\hat{\mathbf{Q}}$, and the rest by zero or identity matrices. As a result, the system

$$\mathbf{Hu} - \mathbf{Gq} = (\mathbf{H}\hat{\mathbf{U}} - \mathbf{G}\hat{\mathbf{Q}})\mathbf{F}^{-1}\mathbf{b} \tag{2.66}$$

is of size $(N + L) \times (N + L)$. After the insertion of boundary conditions, its reduced form $\mathbf{Ax} = \mathbf{b}$ gives both the unknown boundary u and q values and the interior u values at one stroke.

The solution $\mathbf{x}$ of the system (2.66) can be obtained for the solution vector $\mathbf{u}$ and normal derivative vector $\mathbf{q}$ values if the system consists of linear equations, that is, if the right-hand side vector $\mathbf{b}$ is known. This corresponds to the case when the right-hand side function b is a known function of position as $b = b(x, y)$.

When the function b also contains the unknown solution u, its space derivatives $\partial u/\partial x$, $\partial u/\partial y$, and time derivative $\partial u/\partial t$, even appearing in nonlinear form, the system (2.66) needs to be rearranged for the purpose of obtaining the unknowns u and q. Thus, we proceed by taking the right-hand side function $b = b(x, y, u)$, $b = b(x, y, u, u_x, u_y)$, and then $b = b(x, y, t, u, u_x, u_y, u_t)$. For each of these cases, only the vector $\boldsymbol{\alpha} = \mathbf{F}^{-1}\mathbf{b}$ will be changed in Eq. (2.66). The treatment of nonlinear terms and the time derivative will be discussed in detail as is given in [3].

2.2.2 Inhomogeneity Involving the Solution u

First, the application of the dual reciprocity boundary element method is shown on the inhomogeneity term $b = b(x, y, u)$ in the Poisson equation $\nabla^2 u = b$. The function b may also be a combination, sum, or product of functions containing the solution u. This includes linear as well as nonlinear terms. In order to generate the DRBEM discretized equations, the case

$$\nabla^2 u + u = 0 \tag{2.67}$$

is considered as an example as in [3]. Now, the function b becomes $b = -u$ and the vector $\boldsymbol{\alpha}$

$$\boldsymbol{\alpha} = \mathbf{F}^{-1}\mathbf{b} = -\mathbf{F}^{-1}\mathbf{u}\,. \tag{2.68}$$

Then, the DRBEM equation (2.66) takes the form

$$\mathbf{Hu} - \mathbf{Gq} = -(\mathbf{H}\hat{\mathbf{U}} - \mathbf{G}\hat{\mathbf{Q}})\mathbf{F}^{-1}\mathbf{u} \tag{2.69}$$

since the vector $\boldsymbol{\alpha}$ cannot be obtained explicitly. Rewriting equation (2.69) as

$$\mathbf{Hu} - \mathbf{Gq} = -\mathbf{Su} \quad \text{and} \quad (\mathbf{H} + \mathbf{S})\mathbf{u} = \mathbf{Gq} \tag{2.70}$$

with

$$\mathbf{S} = (\mathbf{H}\hat{\mathbf{U}} - \mathbf{G}\hat{\mathbf{Q}})\mathbf{F}^{-1}\,. \tag{2.71}$$

On the boundary, N entries of $\mathbf{u}$ or $\mathbf{q}$ are unknown, while the L values of $\mathbf{u}$ at the interior points are all unknown. The entries of the vector $\mathbf{q}$ from $N+1$ to $N+L$ are multiplied by zero partitions of the matrix $\mathbf{G}$ since $\mathbf{q}$ is not defined there.

Insertion of boundary conditions to Eq. (2.70) results in the usual linear system

$$\mathbf{Ax} = \mathbf{y}, \tag{2.72}$$

which is of size $(N + L) \times (N + L)$, and the $(N + L) \times 1$ vector $\mathbf{x}$ contains N boundary values of $\mathbf{u}$ or $\mathbf{q}$ plus L interior $\mathbf{u}$ values.

2.2.3 Inhomogeneity Involving Convective Terms

Convective terms can easily be treated in the DRBEM formulation by approximating the space derivatives of the solution also using radial basis functions.

We consider an equation of the type [3]

$$\nabla^2 u = \frac{\partial u}{\partial x} + \frac{\partial u}{\partial y} \tag{2.73}$$

in which $b = \partial u/\partial x + \partial u/\partial y = b(x, y, u_x, u_y)$. Thus, the vector $\boldsymbol{\alpha}$ becomes

$$\boldsymbol{\alpha} = \mathbf{F}^{-1}(\frac{\partial \mathbf{u}}{\partial x} + \frac{\partial \mathbf{u}}{\partial y}), \tag{2.74}$$

where $\partial \mathbf{u}/\partial x$ and $\partial \mathbf{u}/\partial y$ are vectors containing the nodal values of x- and y-derivatives of u.

Now, the DRBEM system of Eq. (2.66) becomes

$$\mathbf{Hu} - \mathbf{Gq} = \mathbf{S}(\frac{\partial \mathbf{u}}{\partial x} + \frac{\partial \mathbf{u}}{\partial y}) \,. \tag{2.75}$$

A relationship with the nodal values of u and $\partial u/\partial x$, $\partial u/\partial y$ is established by approximating the solution u itself with the use of radial basis functions f_j and obtaining the system

$$\mathbf{u} = \mathbf{F}\boldsymbol{\beta}, \tag{2.76}$$

where $\boldsymbol{\beta} \neq \boldsymbol{\alpha}$. Differentiating $\mathbf{u}$ with respect to x and y

$$\frac{\partial \mathbf{u}}{\partial x} = \frac{\partial \mathbf{F}}{\partial x}\boldsymbol{\beta}, \quad \frac{\partial \mathbf{u}}{\partial y} = \frac{\partial \mathbf{F}}{\partial y}\boldsymbol{\beta} \tag{2.77}$$

and substituting $\boldsymbol{\beta} = \mathbf{F}^{-1}\mathbf{u}$, the space derivatives of $\mathbf{u}$ can be approximated with the coordinate matrix $\mathbf{F}$ as

$$\frac{\partial \mathbf{u}}{\partial x} = \frac{\partial \mathbf{F}}{\partial x}\mathbf{F}^{-1}\mathbf{u}, \quad \frac{\partial \mathbf{u}}{\partial y} = \frac{\partial \mathbf{F}}{\partial y}\mathbf{F}^{-1}\mathbf{u}\,. \tag{2.78}$$

Then,

$$\mathbf{Hu} - \mathbf{Gq} = \mathbf{S}(\frac{\partial \mathbf{F}}{\partial x}\mathbf{F}^{-1} + \frac{\partial \mathbf{F}}{\partial y}\mathbf{F}^{-1})\mathbf{u} \tag{2.79}$$

results in the system

$$(\mathbf{H} - \mathbf{R})\mathbf{u} = \mathbf{Gq}, \tag{2.80}$$

where

$$\mathbf{R} = \mathbf{S}(\frac{\partial \mathbf{F}}{\partial x}\mathbf{F}^{-1} + \frac{\partial \mathbf{F}}{\partial y}\mathbf{F}^{-1}) \tag{2.81}$$

with the same matrix $\mathbf{S}$ given in Eq. (2.71).

The matrices $\partial \mathbf{F}/\partial x$ and $\partial \mathbf{F}/\partial y$ are of size $(N + L) \times (N + L)$ and obtained by differentiating

$$\begin{aligned}\frac{\partial f(r)}{\partial x} &= \frac{\partial f}{\partial r}\frac{\partial r}{\partial x} = \frac{\partial f}{\partial r}\frac{r_x}{r} = \frac{r_x}{r} + 2r_x + 3r_x r + \cdots + m r_x r^{m-2} \\ \frac{\partial f(r)}{\partial y} &= \frac{\partial f}{\partial r}\frac{\partial r}{\partial y} = \frac{\partial f}{\partial r}\frac{r_y}{r} = \frac{r_y}{r} + 2r_y + 3r_y r + \cdots + m r_y r^{m-2}\,.\end{aligned} \tag{2.82}$$

Second-order space derivatives of the solution vector $\mathbf{u}$ can also be approximated in a similar manner with

$$\frac{\partial^2 \mathbf{u}}{\partial x^2} = \frac{\partial^2 \mathbf{F}}{\partial x^2}\mathbf{F}^{-1}u, \quad \frac{\partial^2 \mathbf{u}}{\partial y^2} = \frac{\partial^2 \mathbf{F}}{\partial y^2}\mathbf{F}^{-1}\mathbf{u} \tag{2.83}$$

if the right-hand side function b involves the second derivatives of u in the Poisson equation $\nabla^2 u = b$.

The nonlinear case $u\frac{\partial u}{\partial x} + u\frac{\partial u}{\partial y}$ of the inhomogeneity b in $\nabla^2 u = b$ can be handled by forming the vector $\mathbf{z} = \mathbf{U}(\frac{\partial \mathbf{F}}{\partial x}\mathbf{F}^{-1} + \frac{\partial \mathbf{F}}{\partial y}\mathbf{F}^{-1})\mathbf{u}$ in $\mathbf{Hu} - \mathbf{Gq} = -\mathbf{Sz}$ as

$$\mathbf{Hu} - \mathbf{Gq} = -(\mathbf{H}\hat{\mathbf{U}} - \mathbf{G}\hat{\mathbf{Q}})\mathbf{F}^{-1}\mathbf{U}(\frac{\partial \mathbf{F}}{\partial x}\mathbf{F}^{-1} + \frac{\partial \mathbf{F}}{\partial y}\mathbf{F}^{-1})\mathbf{u}, \tag{2.84}$$

where $\mathbf{U}$ is a diagonal matrix containing the values of u at the $N + L$ nodes.

2.2.4 Inhomogeneity Involving the Time Derivative

In this section, the application of DRBEM will be presented for transient problems.

As a well-known problem, the diffusion equation is used for showing the DRBEM formulation as given in [3].

Thus, the inhomogeneity in the Poisson equation $\nabla^2 u = b$ is $b(x, y, u_t) = \frac{1}{k}\frac{\partial u}{\partial t}$ for the heat equation

$$\nabla^2 u = \frac{1}{k}\frac{\partial u}{\partial t} \tag{2.85}$$

with appropriate boundary and initial conditions. The time derivative of u is again approximated by radial basis functions f_j assuming the coefficients α_j as a function of time

$$\frac{\partial u}{\partial t}(x, y, t) = \sum_{j=1}^{N+L} f_j(x, y)\alpha_j(t), \tag{2.86}$$

thus giving the system

$$\frac{\partial \mathbf{u}}{\partial t} = \mathbf{F}\boldsymbol{\alpha} \quad \text{and} \quad \boldsymbol{\alpha} = \mathbf{F}^{-1}\frac{\partial \mathbf{u}}{\partial t}. \tag{2.87}$$

Then, the DRBEM equation (2.66) becomes

$$\mathbf{Hu} - \mathbf{Gq} = \frac{1}{k}(\mathbf{H}\hat{\mathbf{U}} - \mathbf{G}\hat{\mathbf{Q}})\mathbf{F}^{-1}\frac{\partial \mathbf{u}}{\partial t} \tag{2.88}$$

and

$$\mathbf{C}\frac{\partial \mathbf{u}}{\partial t} + \mathbf{Hu} = \mathbf{Gq}, \tag{2.89}$$

where

$$\mathbf{C} = -\frac{1}{k}(\mathbf{H}\hat{\mathbf{U}} - \mathbf{G}\hat{\mathbf{Q}})\mathbf{F}^{-1}. \tag{2.90}$$

A standard time-integration scheme can be used as in [3], taking linear approximations for u and q from the time levels t_n and t_{n+1} as

$$u = (1 - \theta_u)u^n + \theta_u u^{n+1}, \quad q = (1 - \theta_q)q^n + \theta_q q^{n+1} \tag{2.91}$$

for the forward difference of $\partial u/\partial t$

$$\frac{\partial u}{\partial t} = \frac{u^{n+1} - u^n}{\Delta t}, \tag{2.92}$$

where $0 < \theta_u, \theta_q < 1$ are relaxation parameters.

Then, Eq. (2.89) is solved iteratively for the transient levels t_{n+1} ($n = 0, 1, 2, \dots$) as

$$(\frac{1}{\Delta t}\mathbf{C} + \theta_u \mathbf{H})\mathbf{u}^{n+1} - \theta_q \mathbf{G}\mathbf{q}^{n+1} = (\frac{1}{\Delta t}\mathbf{C} - (1-\theta_u)\mathbf{H})\mathbf{u}^{n} + (1-\theta_q)\mathbf{G}\mathbf{q}^{n} \qquad (2.93)$$

with given initial condition $u(x, y, t_0) = u_0(x, y)$ and $q(x, y, t_0) = 0$.

Equation (2.93) is a basic DRBEM formulation employing a simple two-level time-integration scheme for the diffusion equation (2.85). Alternative procedures have been used by Gümgüm [36], Bozkaya [37], Singh and Kalra [38], and Lahrmann and Haberland [39].

Gümgüm [36] used the central difference scheme

$$\frac{\partial \mathbf{u}}{\partial t} = \frac{\mathbf{u}^{n+1} - \mathbf{u}^{n-1}}{2\Delta t}$$

for the time derivative in Eq. (2.89). The use of central difference with relaxation parameters θ_u and θ_q for $\mathbf{u}$ and $\partial \mathbf{u}/\partial n$, respectively, gives the DRBEM discretized equation

$$(\frac{1}{2\Delta t}\mathbf{C} + \theta_u \mathbf{H})\mathbf{u}^{n+1} - \theta_q \mathbf{G}\mathbf{q}^{n+1} = (\frac{1}{2\Delta t}\mathbf{C} - (1-\theta_u)\mathbf{H})\mathbf{u}^{n-1} + (1-\theta_q)\mathbf{G}\mathbf{q}^{n-1}, \qquad (2.94)$$

which needs known information from two previous time levels for obtaining the unknown u and q nodal values at the required time level $(n + 1)$.

In Bozkaya [37], the time derivative term in the DRBEM formulation of convection–diffusion equation was discretized by using the differential quadrature method. This gave the solution at all the transient levels directly without the need of a time iteration. Singh and Kalra [38] adopted a least-squares formulation that constructs a functional given by the integral of the square of the error over a time step. A weighted time-step solution was derived by Lahrmann and Haberland [39] to optimize the parameters θ_u and θ_q in Eq. (2.91) that depend on element size, time step, and thermal diffusivity.

2.2.5 *Nonlinearities in the Equations: Navier–Stokes Equations*

The nonlinear terms in the governing equations can be easily incorporated into the DRBEM analysis. Navier–Stokes equations have the structure of a convection–diffusion equation but contain the full nonlinearity of the viscous, incompressible fluid flow equations. In two-dimensional steady-state situations, their usual nondimensional forms are

$$\frac{\partial u}{\partial x} + \frac{\partial v}{\partial y} = 0 \tag{2.95}$$

$$\nabla^2 u = \frac{\partial p}{\partial x} + Re(u\frac{\partial u}{\partial x} + v\frac{\partial u}{\partial y}) \tag{2.96}$$

$$\nabla^2 v = \frac{\partial p}{\partial y} + Re(u\frac{\partial v}{\partial x} + v\frac{\partial v}{\partial y}), \tag{2.97}$$

where (u, v) is the velocity field, p is the pressure, and the parameter Re denotes the Reynolds number. From the momentum equations (2.96)–(2.97), the pressure correction equation [40]

$$\nabla^2 p = 2Re(\frac{\partial u}{\partial x}\frac{\partial v}{\partial y} - \frac{\partial u}{\partial y}\frac{\partial v}{\partial x}) \tag{2.98}$$

can be obtained with the satisfaction of the continuity equation (2.95).

For using the fundamental solution of simpler equation (i.e., Laplace equation), the momentum equations and the pressure correction equation are put into the forms

$$\nabla^2 u = b_1(x, y, u, v, \frac{\partial p}{\partial x}, \frac{\partial u}{\partial x}, \frac{\partial u}{\partial y}) \tag{2.99}$$

$$\nabla^2 v = b_2(x, y, u, v, \frac{\partial p}{\partial y}, \frac{\partial v}{\partial x}, \frac{\partial v}{\partial y}) \tag{2.100}$$

$$\nabla^2 p = b_3(x, y, \frac{\partial u}{\partial x}, \frac{\partial u}{\partial y}, \frac{\partial v}{\partial x}, \frac{\partial v}{\partial y}) . \tag{2.101}$$

One notes that b_1 and b_2 contain the products of the velocity components and their space derivatives, but b_3 has the product of space derivatives of the velocities, respectively. Thus, the $\boldsymbol{\alpha}$ vector in Eq. (2.62) becomes $\boldsymbol{\alpha}_1 = \mathbf{F}^{-1}\mathbf{b}_1$, $\boldsymbol{\alpha}_2 = \mathbf{F}^{-1}\mathbf{b}_2$, and $\boldsymbol{\alpha}_3 = \mathbf{F}^{-1}\mathbf{b}_3$.

The space derivatives are treated as given in Sect. 2.2.3 for obtaining the vectors

$$\frac{\partial \mathbf{u}}{\partial x} = \frac{\partial \mathbf{F}}{\partial x}\mathbf{F}^{-1}\mathbf{u}, \quad \frac{\partial \mathbf{u}}{\partial y} = \frac{\partial \mathbf{F}}{\partial y}\mathbf{F}^{-1}\mathbf{u}, \tag{2.102}$$

$$\frac{\partial \mathbf{v}}{\partial x} = \frac{\partial \mathbf{F}}{\partial x}\mathbf{F}^{-1}\mathbf{v}, \quad \frac{\partial \mathbf{v}}{\partial y} = \frac{\partial \mathbf{F}}{\partial y}\mathbf{F}^{-1}\mathbf{v}, \tag{2.103}$$

$$\frac{\partial \mathbf{p}}{\partial x} = \frac{\partial \mathbf{F}}{\partial x}\mathbf{F}^{-1}\mathbf{p}, \quad \frac{\partial \mathbf{p}}{\partial y} = \frac{\partial \mathbf{F}}{\partial y}\mathbf{F}^{-1}\mathbf{p} . \tag{2.104}$$

Thus, the right-hand side vectors for Eqs. (2.99)–(2.101) through Eqs. (2.96)–(2.98) become

$$\mathbf{b_1} = \frac{\partial \mathbf{F}}{\partial x}\mathbf{F}^{-1}\mathbf{p} + Re(\mathbf{U}\frac{\partial \mathbf{F}}{\partial x} + \mathbf{V}\frac{\partial \mathbf{F}}{\partial y})\mathbf{F}^{-1}\mathbf{u}, \quad (2.105)$$

$$\mathbf{b_2} = \frac{\partial \mathbf{F}}{\partial y}\mathbf{F}^{-1}\mathbf{p} + Re(\mathbf{U}\frac{\partial \mathbf{F}}{\partial x} + \mathbf{V}\frac{\partial \mathbf{F}}{\partial y})\mathbf{F}^{-1}\mathbf{v}, \quad (2.106)$$

$$\mathbf{b_3} = 2Re(\frac{\partial \mathbf{F}}{\partial x}\mathbf{F}^{-1}\mathbf{U}\frac{\partial \mathbf{F}}{\partial y} - \frac{\partial \mathbf{F}}{\partial y}\mathbf{F}^{-1}\mathbf{U}\frac{\partial \mathbf{F}}{\partial x})\mathbf{F}^{-1}\mathbf{v}, \quad (2.107)$$

where $\mathbf{U}$ and $\mathbf{V}$ are diagonal matrices containing u, v nodal values, respectively.

Substituting these $\mathbf{b_1}, \mathbf{b_2}, \mathbf{b_3}$ into the vectors $\boldsymbol{\alpha}_1, \boldsymbol{\alpha}_2, \boldsymbol{\alpha}_3$, respectively, one obtains the DRBEM discretized equations [40]

$$\mathbf{Hu} - \mathbf{Gq}_u = (\mathbf{H}\hat{\mathbf{U}} - \mathbf{G}\hat{\mathbf{Q}})\mathbf{F}^{-1}\mathbf{b}_1, \quad (2.108)$$

$$\mathbf{Hv} - \mathbf{Gq}_v = (\mathbf{H}\hat{\mathbf{U}} - \mathbf{G}\hat{\mathbf{Q}})\mathbf{F}^{-1}\mathbf{b}_2, \quad (2.109)$$

$$\mathbf{Hp} - \mathbf{Gq}_p = (\mathbf{H}\hat{\mathbf{U}} - \mathbf{G}\hat{\mathbf{Q}})\mathbf{F}^{-1}\mathbf{b}_3, \quad (2.110)$$

where $\mathbf{q}_u, \mathbf{q}_v, \mathbf{q}_p$ denote the vectors of the normal derivatives of u, v, and p nodal values, respectively.

Defining matrices $\mathbf{S}, \mathbf{S}_1, \mathbf{S}_2, \mathbf{R}$, and $\mathbf{T}$ as

$$\mathbf{S} = (\mathbf{H}\hat{\mathbf{U}} - \mathbf{G}\hat{\mathbf{Q}})\mathbf{F}^{-1}, \;\; \mathbf{S}_1 = \mathbf{S}\frac{\partial \mathbf{F}}{\partial x}\mathbf{F}^{-1}, \;\; \mathbf{S}_2 = \mathbf{S}\frac{\partial \mathbf{F}}{\partial y}\mathbf{F}^{-1},$$

$$\mathbf{R} = \mathbf{S}(\mathbf{U}\frac{\partial \mathbf{F}}{\partial x} + \mathbf{V}\frac{\partial \mathbf{F}}{\partial y})\mathbf{F}^{-1} \quad (2.111)$$

$$\mathbf{T} = \mathbf{S}(\frac{\partial \mathbf{F}}{\partial x}\mathbf{F}^{-1}\mathbf{U}\frac{\partial \mathbf{F}}{\partial y} - \frac{\partial \mathbf{F}}{\partial y}\mathbf{F}^{-1}\mathbf{U}\frac{\partial \mathbf{F}}{\partial x})\mathbf{F}^{-1},$$

Equations (2.108)–(2.110) become

$$(\mathbf{H} - Re\mathbf{R})\mathbf{u} = \mathbf{Gq}_u + \mathbf{S}_1\mathbf{p}, \quad (2.112)$$

$$(\mathbf{H} - Re\mathbf{R})\mathbf{v} = \mathbf{Gq}_v + \mathbf{S}_2\mathbf{p}, \quad (2.113)$$

$$\mathbf{Hp} - \mathbf{Gq}_p = 2Re\mathbf{Tv}, \quad (2.114)$$

where the matrices $\mathbf{R}$ and $\mathbf{T}$ are functions of u and v nodal values.

The solution procedure is now iterative due to the nonlinear terms of the momentum and pressure correction equations. An iterative algorithm is used as:

- Supply initial values of **u**, **v**, **p**.
- Solve x-momentum equation (2.112) for **u**.
- Solve y-momentum equation (2.113) for **v** using updated **u** values.
- Solve pressure equation (2.114) for **p** using the latest values of **u** and **v**.
- Continue iteration until convergence for a preassigned tolerance.

Both of the BEM and DRBEM formulations expressed for the Poisson type equations are going to be used in solving MHD flow problems in the rest of the book. The MHD duct flow problems, MHD flow problems in infinite channels and in infinite regions will be solved by direct BEM application using the fundamental solution for coupled MHD equations, which will be derived in Chap. 3. Then, the unsteady MHD duct flow problems, MHD flow equations in the case of low magnetic Reynolds number, R_m, and natural convection MHD flows with or without induced magnetic field, MHD flows involving electric potential are going to be treated with the DRBEM procedure. The reason is that the equations involve time derivatives and/or some nonlinear terms in which the fundamental solutions are not available.

The solution of the resulting linear system of equations in BEM and DRBEM depends on the coefficient matrix of the system that has some drawbacks. It is fully populated with the nonzero entries and not symmetric. This means that the entire BEM coefficient matrix must be saved in the computer core memory.

In the MHD flow problems solved by BEM (Chap. 4) and DRBEM (Chap. 5), the resulting system of equations are solved by using the solver *mldivide* in MATLAB that returns to the LU decomposition of the coefficient matrix in square systems and the least-squares solution for rectangular systems. Since only the boundary of the region is discretized, the size of the BEM systems is not too large compared to the FDM and FEM resulted systems. Thus, the solver *mldivide* does not give a problem in obtaining accurate results of the considered problems. The programs are written in MATLAB language.

In the literature some preconditioned algorithms were found to work quite well in solving BEM or DRBEM system of equations. Mansur et.al. [41] have been solved BEM system of equations via iterative techniques. A linear system solver has been proposed by adopting the product type method as a preconditioner to the generalized minimal residual method (GMRES) in [42]. The solver has a faster speed to the Gaussian elimination when BEM is applied to Laplace equation. Araújo et.al. [43] have been constructed global block-diagonal preconditioners. In Wathen et.al. [44], a preconditioning technique that exploits the block structure of linear system resulting from mixed FEM solution of MHD equations has been given especially for fine mesh discretization (large-size problems). The results of the MHD problems considered in this book are obtained by solving the resulting BEM and DRBEM linear system of equations with the MATLAB solver *mldivide* without the need of preconditioning.

Chapter 3
Fundamental Solution to Coupled MHD Flow Equations

In the present chapter, the application of the boundary element method to the steady equations of the MHD duct flow, MHD flow in infinite regions, and MHD flow between infinite parallel plates is considered with a fundamental solution that treats the equations in their coupled form. The direct BEM approach requires the inherent use of the fundamental solution to the governing equations. Thus, Sect. 3.1 is devoted to the derivation of the fundamental solution for the steady form of the MHD flow equations (1.57) in the original coupled form that are convection–diffusion type. Horizontally and vertically applied magnetic field cases (Eqs. (1.61) and (1.62)) and electrically driven MHD flow equations (1.63) are all tackled with this fundamental solution. Then, in Sect. 3.2, the application of the boundary element method to these coupled MHD flow equations by the use of the fundamental solution derived in Sect. 3.1 is explained. Thus, the dimensionality of the problem is reduced by one, and the constant elements are used in order to discretize the boundary of the domain under consideration. The discretized BEM system of linear equations is obtained, and the boundary conditions have been accounted for. The details regarding the derivation of the fundamental solution and the application of the BEM to the steady MHD flow equations can also be found in the work of Bozkaya and Tezer-Sezgin [45][1] and Bozkaya [37].

[1] A summary of Chap. 3 was published in Journal of Computational and Applied Mathematics, Volume 203(1), C. Bozkaya and M. Tezer-Sezgin, Fundamental solution for coupled magnetohydrodynamic flow equations, 125–144, Copyright Elsevier (2007).

M. Tezer-Sezgin, C. Bozkaya, *Boundary Element Method for Magnetohydrodynamic Flow*, Surveys and Tutorials in the Applied Mathematical Sciences 14,
https://doi.org/10.1007/978-3-031-58353-7_3

3.1 Fundamental Solution of Coupled MHD Flow Equations

The steady form of the governing MHD flow equations under the effect of an externally applied inclined magnetic field given in Eq. (1.57) can be written as follows:

$$\begin{aligned} \nabla^2 V + Ha_x \frac{\partial B}{\partial x} + Ha_y \frac{\partial B}{\partial y} &= -1 \\ \nabla^2 B + Ha_x \frac{\partial V}{\partial x} + Ha_y \frac{\partial V}{\partial y} &= 0 \end{aligned} , \tag{3.1}$$

which describes two coupled second-order linear partial differential equations. The region is either a two-dimensional duct or an infinite plane $y \geq 0$ or an infinite region between parallel plates. The boundary $\Gamma = \Gamma_1 \cup \Gamma_2$ is assumed to be partly insulated–partly perfectly conducting, whereas the velocity is also taken as no-slip. For the BEM solution of Eqs. (3.1) that are the convection–diffusion type coupled equations, the fundamental solution is needed because the BEM is a numerical technique that makes intensive use of a fundamental solution of the problem in question.

The MHD equations (3.1) are first transformed into the matrix–vector form

$$\mathbf{L}\,\mathbf{u} = \mathbf{f}, \tag{3.2}$$

where $\mathbf{L}$ is the matrix operator containing both diffusion and convection operators

$$\mathbf{L} = \begin{bmatrix} \nabla^2 & Ha_x \frac{\partial}{\partial x} + Ha_y \frac{\partial}{\partial y} \\ Ha_x \frac{\partial}{\partial x} + Ha_y \frac{\partial}{\partial y} & \nabla^2 \end{bmatrix} \tag{3.3}$$

and

$$\mathbf{u} = \begin{bmatrix} V \\ B \end{bmatrix}, \quad \mathbf{f} = \begin{bmatrix} -1 \\ 0 \end{bmatrix}. \tag{3.4}$$

Weighting this equation over the domain of the problem, Ω, in the weighted residuals principle, [2]

$$\int_\Omega \mathbf{w}^T \mathbf{L}\,\mathbf{u}\, d\Omega = \int_\Omega \mathbf{w}^T \mathbf{f}\, d\Omega, \tag{3.5}$$

where $\mathbf{w}^T$ is the transpose of the vector weight function $\mathbf{w} = \begin{bmatrix} V^* \\ B^* \end{bmatrix}$, gives the integral equation

$$\int_{\Omega} \left[V^* \; B^* \right] \begin{bmatrix} \nabla^2 & Ha_x \dfrac{\partial}{\partial x} + Ha_y \dfrac{\partial}{\partial y} \\ Ha_x \dfrac{\partial}{\partial x} + Ha_y \dfrac{\partial}{\partial y} & \nabla^2 \end{bmatrix} \begin{bmatrix} V \\ B \end{bmatrix} d\Omega$$

$$= \int_{\Omega} \left[V^* \; B^* \right] \begin{bmatrix} -1 \\ 0 \end{bmatrix} d\Omega \; .$$

This is actually one equation containing three integrals in the form

$$I_1 + I_2 = I_3, \tag{3.6}$$

where

$$I_1 = \int_{\Omega} V^* \left(\nabla^2 V + Ha_x \frac{\partial B}{\partial x} + Ha_y \frac{\partial B}{\partial y} \right) d\Omega$$

$$I_2 = \int_{\Omega} B^* \left(\nabla^2 B + Ha_x \frac{\partial V}{\partial x} + Ha_y \frac{\partial V}{\partial y} \right) d\Omega$$

and

$$I_3 = -\int_{\Omega} V^* d\Omega \; .$$

After the application of Green's second identity, the integrals I_1 and I_2 reduce to

$$I_1 = \int_{\Omega} V \nabla^2 V^* d\Omega + \int_{\Gamma} \left(V^* \frac{\partial V}{\partial n} - V \frac{\partial V^*}{\partial n} \right) d\Gamma + I_4 \tag{3.7}$$

$$I_2 = \int_{\Omega} B \nabla^2 B^* d\Omega + \int_{\Gamma} \left(B^* \frac{\partial B}{\partial n} - B \frac{\partial B^*}{\partial n} \right) d\Gamma + I_5 \tag{3.8}$$

where Γ is the boundary of the domain Ω and

$$I_4 = \int_{\Omega} V^* \left(Ha_x \frac{\partial B}{\partial x} + Ha_y \frac{\partial B}{\partial y} \right) d\Omega$$

and

$$I_5 = \int_{\Omega} B^* \left(Ha_x \frac{\partial V}{\partial x} + Ha_y \frac{\partial V}{\partial y} \right) d\Omega \; .$$

Now, we rewrite

$$
\begin{aligned}
V^*\left(Ha_x\frac{\partial B}{\partial x}+Ha_y\frac{\partial B}{\partial y}\right) &= Ha_x\left(\left(V^*B\right)_{,x}-V^*_{,x}B\right)\\
&+Ha_y\left(\left(V^*B\right)_{,y}-V^*_{,y}B\right),\\
B^*\left(Ha_x\frac{\partial V}{\partial x}+Ha_y\frac{\partial V}{\partial y}\right) &= Ha_x\left(\left(B^*V\right)_{,x}-B^*_{,x}V\right)\\
&+Ha_y\left(\left(B^*V\right)_{,y}-B^*_{,y}V\right)
\end{aligned}
$$

and substitute in the integrals I_4 and I_5. The subscripts $_{,x}$, $_{,y}$ denote the x and y derivatives, respectively. Moreover, with the application of the Green's theorem, the integrals I_4 and I_5 become

$$
\begin{aligned}
I_4 &= \int_\Gamma \left(Ha_xV^*Bn_x+Ha_yV^*Bn_y\right)d\Gamma-\int_\Omega\left(Ha_xBV^*_{,x}+Ha_yBV^*_{,y}\right)d\Omega\\
I_5 &= \int_\Gamma \left(Ha_xB^*Vn_x+Ha_yB^*Vn_y\right)d\Gamma-\int_\Omega\left(Ha_xVB^*_{,x}+Ha_yVB^*_{,y}\right)d\Omega\ .
\end{aligned}
$$

By the substitution of I_4 and I_5 in Eqs. (3.7) and (3.8), respectively, Eq. (3.6) takes the form

$$
\begin{aligned}
&\int_\Omega V\left(\nabla^2V^*-Ha_x\frac{\partial B^*}{\partial x}-Ha_y\frac{\partial B^*}{\partial y}\right)d\Omega\\
&+\int_\Omega B\left(\nabla^2B^*-Ha_x\frac{\partial V^*}{\partial x}-Ha_y\frac{\partial V^*}{\partial y}\right)d\Omega\\
&+\int_\Gamma\left(V^*\frac{\partial V}{\partial n}-V\frac{\partial V^*}{\partial n}\right)d\Gamma+\int_\Gamma\left(B^*\frac{\partial B}{\partial n}-B\frac{\partial B^*}{\partial n}\right)d\Gamma\ ,\\
&+\int_\Gamma Ha_x\left(V^*B+B^*V\right)n_xd\Gamma+\int_\Gamma Ha_y\left(V^*B+B^*V\right)n_yd\Gamma\\
&=-\int_\Omega V^*d\Omega
\end{aligned}
\tag{3.9}
$$

where $\mathbf{n}=(n_x,n_y)$ is the outward unit normal vector on Γ.

To omit the region integrals on the left-hand side of Eq. (3.9), we need to consider two cases:

First Case

$$\begin{aligned} \nabla^2 V^* - Ha_x \frac{\partial B^*}{\partial x} - Ha_y \frac{\partial B^*}{\partial y} &= -\Delta_A(P) \\ \nabla^2 B^* - Ha_x \frac{\partial V^*}{\partial x} - Ha_y \frac{\partial V^*}{\partial y} &= 0 \end{aligned},$$

where A and P are the fixed (source) and variable (field) points in Ω, respectively, and Δ_A is the Dirac delta function at the source point A. The solution for the equations in the first case is denoted as

$$\begin{bmatrix} V_1^* \\ B_1^* \end{bmatrix}.$$

Second Case

$$\begin{aligned} \nabla^2 V^* - Ha_x \frac{\partial B^*}{\partial x} - Ha_y \frac{\partial B^*}{\partial y} &= 0 \\ \nabla^2 B^* - Ha_x \frac{\partial V^*}{\partial x} - Ha_y \frac{\partial V^*}{\partial y} &= -\Delta_A(P) \end{aligned},$$

which we denote the solution as

$$\begin{bmatrix} V_2^* \\ B_2^* \end{bmatrix}.$$

These two cases transform Eq. (3.9) into the following two integral equations:

$$\begin{aligned} -c_A V(A) &+ \int_\Gamma \left(V_1^* \frac{\partial V}{\partial n} - V \frac{\partial V_1^*}{\partial n} \right) d\Gamma + \int_\Gamma \left(B_1^* \frac{\partial B}{\partial n} - B \frac{\partial B_1^*}{\partial n} \right) d\Gamma \\ &+ \int_\Gamma Ha_x \left(V_1^* B + B_1^* V \right) n_x d\Gamma + \int_\Gamma Ha_y \left(V_1^* B + B_1^* V \right) n_y d\Gamma \\ &= -\int_\Omega V_1^* d\Omega \end{aligned} \tag{3.10}$$

and

$$
\begin{aligned}
-c_A B(A) &+ \int_\Gamma \left(V_2^* \frac{\partial V}{\partial n} - V \frac{\partial V_2^*}{\partial n} \right) d\Gamma + \int_\Gamma \left(B_2^* \frac{\partial B}{\partial n} - B \frac{\partial B_2^*}{\partial n} \right) d\Gamma \\
&+ \int_\Gamma Ha_x \left(V_2^* B + B_2^* V \right) n_x d\Gamma + \int_\Gamma Ha_y \left(V_2^* B + B_2^* V \right) n_y d\Gamma \quad , \qquad (3.11) \\
&= - \int_\Omega V_2^* d\Omega
\end{aligned}
$$

where c_A is a constant equal to $\dfrac{\theta}{2\pi}$, θ being the internal angle at the source point A.

Let the matrix $\mathbf{G}^*$ be formed as

$$
\mathbf{G}^* = \begin{bmatrix} V_1^* & V_2^* \\ B_1^* & B_2^* \end{bmatrix}, \tag{3.12}
$$

which is the fundamental solution for the adjoint operator

$$
\mathbf{L}^* = \begin{bmatrix} \nabla^2 & -Ha_x \dfrac{\partial}{\partial x} - Ha_y \dfrac{\partial}{\partial y} \\ -Ha_x \dfrac{\partial}{\partial x} - Ha_y \dfrac{\partial}{\partial y} & \nabla^2 \end{bmatrix} \tag{3.13}
$$

of $\mathbf{L}$. That is,

$$
\mathbf{L}^* \, \mathbf{G}^* = -\Delta_A(P) \mathbf{I},
$$

and $\mathbf{I}$ is the 2×2 identity matrix since the left-hand side is a 2×2 matrix.

Therefore, the fundamental solution $\mathbf{G}^*$ becomes

$$
\mathbf{G}^* = \begin{bmatrix} \nabla^2 & Ha_x \dfrac{\partial}{\partial x} + Ha_y \dfrac{\partial}{\partial y} \\ Ha_x \dfrac{\partial}{\partial x} + Ha_y \dfrac{\partial}{\partial y} & \nabla^2 \end{bmatrix} \Phi, \tag{3.14}
$$

where Φ is the fundamental solution of the biharmonic equation

$$
\left(\nabla^4 - Ha_x^2 \frac{\partial^2}{\partial x^2} - 2 Ha_x Ha_y \frac{\partial^2}{\partial x \partial y} - Ha_y^2 \frac{\partial^2}{\partial y^2} \right) u = 0 \, .
$$

That is, the fundamental solution Φ satisfies the equation

$$
\left(\nabla^4 - Ha_x^2 \frac{\partial^2}{\partial x^2} - 2 Ha_x Ha_y \frac{\partial^2}{\partial x \partial y} - Ha_y^2 \frac{\partial^2}{\partial y^2} \right) \Phi = -\Delta_A(P) \, .
$$

This can be partitioned into the following two convection–diffusion equations

$$\left(\nabla^2 - Ha_x\frac{\partial}{\partial x} - Ha_y\frac{\partial}{\partial y}\right)\Psi_1 = -\Delta_A(P) \tag{3.15}$$

and

$$\left(\nabla^2 + Ha_x\frac{\partial}{\partial x} + Ha_y\frac{\partial}{\partial y}\right)\Psi_2 = -\Delta_A(P), \tag{3.16}$$

where

$$\Psi_1 = \left(\nabla^2 + Ha_x\frac{\partial}{\partial x} + Ha_y\frac{\partial}{\partial y}\right)\Phi \tag{3.17}$$

and

$$\Psi_2 = \left(\nabla^2 - Ha_x\frac{\partial}{\partial x} - Ha_y\frac{\partial}{\partial y}\right)\Phi\ . \tag{3.18}$$

It is clear that Ψ_1 and Ψ_2 are the fundamental solutions of the convection–diffusion type equations (3.15) and (3.16), respectively. Therefore,

$$\Psi_1 = \frac{1}{2\pi}\ e^{\mathbf{M}\cdot\,\mathbf{r}/2}K_0(\frac{Ha}{2}r) \tag{3.19}$$

and

$$\Psi_2 = \frac{1}{2\pi}\ e^{-\mathbf{M}\cdot\,\mathbf{r}/2}K_0(\frac{Ha}{2}r), \tag{3.20}$$

where $\mathbf{M}$ is the vector with components $\mathbf{M} = (Ha_x, Ha_y)$ and $\mathbf{r} = (r_x, r_y)$ is the distance vector between the source and field points. The Hartmann number (Ha) and r are the modulus of the vectors $\mathbf{M}$ and $\mathbf{r}$, respectively. K_0 is the modified Bessel function of the second kind and of order zero.

With the relationships between Φ and the fundamental solutions Ψ_1 and Ψ_2, we have

$$Ha_x\frac{\partial\Phi}{\partial x} + Ha_y\frac{\partial\Phi}{\partial y} = \frac{\Psi_1 - \Psi_2}{2} \tag{3.21}$$

$$\nabla^2\Phi = \frac{\Psi_1 + \Psi_2}{2}, \tag{3.22}$$

where

$$\frac{\Psi_1 - \Psi_2}{2} = \frac{1}{2\pi} K_0(\frac{Ha}{2} r) \sinh(\frac{\mathbf{M} \cdot \mathbf{r}}{2})$$

and

$$\frac{\Psi_1 + \Psi_2}{2} = \frac{1}{2\pi} K_0(\frac{Ha}{2} r) \cosh(\frac{\mathbf{M} \cdot \mathbf{r}}{2}) .$$

Thus, the fundamental solution $\mathbf{G}^*$ for the adjoint operator $\mathbf{L}^*$ is finally obtained as

$$\mathbf{G}^* = \begin{bmatrix} \frac{1}{2\pi} K_0(\frac{Ha}{2} r) \cosh(\frac{\mathbf{M} \cdot \mathbf{r}}{2}) & \frac{1}{2\pi} K_0(\frac{Ha}{2} r) \sinh(\frac{\mathbf{M} \cdot \mathbf{r}}{2}) \\ \frac{1}{2\pi} K_0(\frac{Ha}{2} r) \sinh(\frac{\mathbf{M} \cdot \mathbf{r}}{2}) & \frac{1}{2\pi} K_0(\frac{Ha}{2} r) \cosh(\frac{\mathbf{M} \cdot \mathbf{r}}{2}) \end{bmatrix} \tag{3.23}$$

with its entries from (3.12) as

$$\begin{aligned} V_1^* = B_2^* &= \frac{1}{2\pi} K_0(\frac{Ha}{2} r) \cosh(\frac{\mathbf{M} \cdot \mathbf{r}}{2}) \\ V_2^* = B_1^* &= \frac{1}{2\pi} K_0(\frac{Ha}{2} r) \sinh(\frac{\mathbf{M} \cdot \mathbf{r}}{2}) \end{aligned} \tag{3.24}$$

and their normal derivatives

$$\begin{aligned} \frac{\partial V_1^*}{\partial n} = \frac{\partial B_2^*}{\partial n} &= \frac{Ha}{4\pi} K_1(\frac{Ha}{2} r) \cosh(\frac{\mathbf{M} \cdot \mathbf{r}}{2}) \frac{\partial r}{\partial n} + \frac{1}{4\pi} K_0(\frac{Ha}{2} r) \sinh(\frac{\mathbf{M} \cdot \mathbf{r}}{2}) \mathbf{M} \cdot \frac{\partial \mathbf{r}}{\partial n} \\ \frac{\partial V_2^*}{\partial n} = \frac{\partial B_1^*}{\partial n} &= \frac{Ha}{4\pi} K_1(\frac{Ha}{2} r) \sinh(\frac{\mathbf{M} \cdot \mathbf{r}}{2}) \frac{\partial r}{\partial n} + \frac{1}{4\pi} K_0(\frac{Ha}{2} r) \cosh(\frac{\mathbf{M} \cdot \mathbf{r}}{2}) \mathbf{M} \cdot \frac{\partial \mathbf{r}}{\partial n} . \end{aligned} \tag{3.25}$$

3.2 Application of the Boundary Element Method

Having found the fundamental solutions $V_1^* = B_2^*$ and $B_1^* = V_2^*$ as in (3.24), Eqs. (3.10) and (3.11) can be rewritten,

$$-c_A V(A) + \int_\Gamma \left(Ha_x B_1^* n_x + Ha_y B_1^* n_y - \frac{\partial V_1^*}{\partial n} \right) V d\Gamma$$
$$+ \int_\Gamma \left(Ha_x V_1^* n_x + Ha_y V_1^* n_y - \frac{\partial B_1^*}{\partial n} \right) B\, d\Gamma + \int_\Gamma V_1^* \frac{\partial V}{\partial n}\, d\Gamma + \int_\Gamma B_1^* \frac{\partial B}{\partial n}\, d\Gamma$$
$$= - \int_\Omega V_1^* d\Omega \tag{3.26}$$

and

$$- c_A B(A) + \int_\Gamma \left(Ha_x V_1^* n_x + Ha_y V_1^* n_y - \frac{\partial B_1^*}{\partial n} \right) V d\Gamma$$
$$+ \int_\Gamma \left(Ha_x B_1^* n_x + Ha_y B_1^* n_y - \frac{\partial V_1^*}{\partial n} \right) B\, d\Gamma + \int_\Gamma B_1^* \frac{\partial V}{\partial n}\, d\Gamma + \int_\Gamma V_1^* \frac{\partial B}{\partial n}\, d\Gamma$$
$$= - \int_\Omega B_1^* d\Omega\,. \tag{3.27}$$

Thus, after the discretization of the boundary Γ of the domain Ω by using constant elements, boundary element matrix equations for the unknowns, the velocity V and the induced magnetic field B and their normal derivatives, can now be obtained through the evaluation of the boundary integrals in Eqs. (3.26) and (3.27). The matrix–vector form is (assembly for all boundary elements)

$$\begin{bmatrix} -c_A \mathbf{V}(A) \\ -c_A \mathbf{B}(A) \end{bmatrix} + \begin{bmatrix} \mathbf{H}\ \mathbf{G} \\ \mathbf{G}\ \mathbf{H} \end{bmatrix} \begin{bmatrix} \mathbf{V} \\ \mathbf{B} \end{bmatrix} + \begin{bmatrix} \mathbf{H}^1\ \mathbf{G}^1 \\ \mathbf{G}^1\ \mathbf{H}^1 \end{bmatrix} \begin{bmatrix} \dfrac{\partial \mathbf{V}}{\partial n} \\ \dfrac{\partial \mathbf{B}}{\partial n} \end{bmatrix} = \begin{bmatrix} \mathbf{F}_1 \\ \mathbf{F}_2 \end{bmatrix}, \tag{3.28}$$

where $\mathbf{H}$, $\mathbf{G}$, $\mathbf{H}^1$, and $\mathbf{G}^1$ are the matrices with the entries

$$\begin{aligned} h_{ij} &= \int_{\Gamma_j} \left(Ha_x B_1^* n_x + Ha_y B_1^* n_y - \frac{\partial V_1^*}{\partial n} \right) d\Gamma_j \\ g_{ij} &= \int_{\Gamma_j} \left(Ha_x V_1^* n_x + Ha_y V_1^* n_y - \frac{\partial B_1^*}{\partial n} \right) d\Gamma_j \\ h_{ij}^1 &= \int_{\Gamma_j} V_1^*\, d\Gamma_j \\ g_{ij}^1 &= \int_{\Gamma_j} B_1^*\, d\Gamma_j \end{aligned} \tag{3.29}$$

where $i, j = 1, \ldots, N$. $\mathbf{F} = \begin{bmatrix} \mathbf{F}_1 \\ \mathbf{F}_2 \end{bmatrix}$ is the right-hand side vector with the entries containing domain integrals

$$\begin{aligned} F_1^i &= -\int_\Omega V_1^* d\Omega \\ &, \\ F_2^i &= -\int_\Omega B_1^* d\Omega \end{aligned} \tag{3.30}$$

which can be calculated by using some numerical integration techniques for $i = 1, \ldots, N$. However, in the applications of BEM to the MHD flow problems in Chap. 4, the nonhomogeneous governing MHD flow equations (3.1) will be transformed into the corresponding homogeneous form by using some suitable particular solutions.

The subscripts i and j indicate the fixed node i and the jth element on the boundary, respectively. $\mathbf{r} = (r_x, r_y)$ is the vector between the boundary nodes i and j. After the substitution of the fundamental solutions V_1^* and B_1^* and their normal derivatives in Eqs. (3.29) and (3.30), the entries of $\mathbf{H}$, $\mathbf{G}$, $\mathbf{H}^1$, $\mathbf{G}^1$, and $\mathbf{F}$ become

$$\begin{aligned}
h_{ij} &= \frac{1}{4\pi}\int_{\Gamma_j}\left(K_0(\frac{Ha}{2}r)\sinh(\frac{\mathbf{M}\cdot\mathbf{r}}{2})\mathbf{M}\cdot\mathbf{n} + Ha\,K_1(\frac{Ha}{2}r)\cosh(\frac{\mathbf{M}\cdot\mathbf{r}}{2})\,\frac{\partial r}{\partial n}\right)d\Gamma_j \\
g_{ij} &= \frac{1}{4\pi}\int_{\Gamma_j}\left(K_0(\frac{Ha}{2}r)\cosh(\frac{\mathbf{M}\cdot\mathbf{r}}{2})\mathbf{M}\cdot\mathbf{n} + Ha\,K_1(\frac{Ha}{2}r)\sinh(\frac{\mathbf{M}\cdot\mathbf{r}}{2})\,\frac{\partial r}{\partial n}\right)d\Gamma_j \\
h_{ij}^1 &= \frac{1}{2\pi}\int_{\Gamma_j} K_0(\frac{Ha}{2}r)\cosh(\frac{\mathbf{M}\cdot\mathbf{r}}{2})\,d\Gamma_j \\
g_{ij}^1 &= \frac{1}{2\pi}\int_{\Gamma_j} K_0(\frac{Ha}{2}r)\sinh(\frac{\mathbf{M}\cdot\mathbf{r}}{2})\,d\Gamma_j \\
F_1^i &= -\int_\Omega K_0(\frac{Ha}{2}r)\cosh(\frac{\mathbf{M}\cdot\mathbf{r}}{2})\,d\Omega \\
F_2^i &= -\int_\Omega K_0(\frac{Ha}{2}r)\sinh(\frac{\mathbf{M}\cdot\mathbf{r}}{2})\,d\Omega
\end{aligned} \tag{3.31}$$

where K_1 is the modified Bessel function of the second kind and of order one.

The constant c_A is $1/2$ or 1 when the fixed point A is on the straight boundary or inside Ω, respectively. The problem is solved first for the values of unknowns V, B and their normal derivatives $\frac{\partial V}{\partial n}$, $\frac{\partial B}{\partial n}$ on the boundary, so Eq. (3.28) becomes

$$\begin{bmatrix} \bar{\mathbf{H}} & \mathbf{G} \\ \mathbf{G} & \bar{\mathbf{H}} \end{bmatrix} \begin{bmatrix} \mathbf{V} \\ \mathbf{B} \end{bmatrix} + \begin{bmatrix} \mathbf{H}^1 & \mathbf{G}^1 \\ \mathbf{G}^1 & \mathbf{H}^1 \end{bmatrix} \begin{bmatrix} \dfrac{\partial \mathbf{V}}{\partial n} \\ \dfrac{\partial \mathbf{B}}{\partial n} \end{bmatrix} = \begin{bmatrix} \mathbf{F}_1 \\ \mathbf{F}_2 \end{bmatrix}, \tag{3.32}$$

where $\bar{\mathbf{H}}$ is the matrix with the entries

$$\bar{h}_{ij} = -\frac{1}{2}\,\delta_{ij} + h_{ij}\ .$$

This linear system of Eq. (3.32) is going to be solved for values of $\frac{\partial \mathbf{V}}{\partial n}$ on Γ, $\mathbf{B}$ on Γ_2, and $\frac{\partial \mathbf{B}}{\partial n}$ on Γ_1 (since $\mathbf{B}$ is given on Γ_1 and $\frac{\partial \mathbf{B}}{\partial n}$ is given on Γ_2). The known boundary values ($\mathbf{V}$ on Γ, $\mathbf{B}$ on Γ_1, and $\frac{\partial \mathbf{B}}{\partial n}$ on Γ_2) are going to be inserted in Eq. (3.32) by shuffling the rows and columns corresponding to the known entries of $[\mathbf{V} \quad \mathbf{B}]^T$ and $[\partial \mathbf{V}/\partial n \quad \partial \mathbf{B}/\partial n]^T$. Having found $(\mathbf{V}, \mathbf{B})$ and $(\frac{\partial \mathbf{V}}{\partial n}, \frac{\partial \mathbf{B}}{\partial n})$ everywhere on the boundary, one can obtain the values of V and B at any point of the domain Ω by using Eq. (3.28) and taking $c_A = 1$. This is given by

$$\begin{bmatrix} \mathbf{V} \\ \mathbf{B} \end{bmatrix} = \begin{bmatrix} \mathbf{HI} & \mathbf{GI} \\ \mathbf{GI} & \mathbf{HI} \end{bmatrix} \begin{bmatrix} \mathbf{V} \\ \mathbf{B} \end{bmatrix} + \begin{bmatrix} \mathbf{HI}^1 & \mathbf{GI}^1 \\ \mathbf{GI}^1 & \mathbf{HI}^1 \end{bmatrix} \begin{bmatrix} \dfrac{\partial \mathbf{V}}{\partial n} \\ \dfrac{\partial \mathbf{B}}{\partial n} \end{bmatrix} - \begin{bmatrix} \mathbf{F}_1 \\ \mathbf{F}_2 \end{bmatrix}, \tag{3.33}$$

where $\mathbf{V}$ and $\mathbf{B}$ are computed at L interior points. Thus, in this computation, $\mathbf{V}$ and $\mathbf{B}$ are of the size $L \times 1$. The entries of the matrices $\mathbf{HI}$, $\mathbf{GI}$, $\mathbf{HI}^1$, and $\mathbf{GI}^1$ are in the same form with the ones of $\mathbf{H}$, $\mathbf{G}$, $\mathbf{H}^1$, and $\mathbf{G}^1$. But, this time in the entries of $\mathbf{HI}$, $\mathbf{GI}$, $\mathbf{HI}^1$, and $\mathbf{GI}^1$, i indicates the fixed node in the domain and j indicates the jth element on the boundary. Thus, the vector $\mathbf{r}$ is computed between the inside node i and the boundary node j.

3.3 Algorithm

A MATLAB computer program created by the authors constructs all the matrices and the right-hand vectors involved in Eqs. (3.32) and (3.33). Its linear solver, namely *mldivide*, is used effectively to derive the solution of the system of linear equations (3.32).

Algorithm 1: Main Code: BEM solution of the MHD flow in a square duct

---

Input:

Problem parameters: Ha: Hartmann number, γ: inclination angle of applied magnetic field with components (Ha_x, Ha_y), ℓ_x, ℓ_y: side length of square duct in x and y-directions

Discretization parameters: $IPx,\ IPy$: number of constant boundary elements in x and y-directions along one wall of the duct, IG: number of Gaussian quadrature nodes

Output: The approximate velocity V and induced magnetic field B on the boundary and interior nodes

1 **Step 1:** Call function **discretization** to discretize the boundary of the duct

2 **Step 2:** Call function **HG-Matrix-Generator** to construct the discretized BEM matrices given in Eqs. (3.32) and (3.33)

3 **Step 3:** Call function **solver** to construct the final linear system of equations obtained after the insertion of no-slip velocity ($V = 0$) and electrically insulated ($B = 0$) boundary conditions in Eq. (3.32). This function returns the approximate BEM solution for V and B.

Combining the results of Sects. 2.1 and 3.1–3.2, Algorithm 1 for the BEM solution of the MHD flow in a duct with no-slip and electrically insulated walls is made up of the following steps. The algorithms for Step 1, Step 2, and Step 3 mentioned in Algorithm 1 are provided at the end of Chap. 3.

Next chapter is going to give numerical solutions for the steady equations of the MHD duct flow, MHD flow between parallel plates, and MHD flow in infinite regions by using the fundamental solution of coupled MHD equations derived in the present chapter. The systems (3.32) and (3.33) are going to be solved imposing no-slip velocity and partly insulated partly conducting induced magnetic field wall conditions.

Algorithm 2: Step 1: Discretization of the computational domain

Input: The problem and discretization parameters given in Main Algorithm 1 as input

Output: The endpoints x, y and midpoints x_m, y_m of boundary elements, numbers of boundary (N) and interior (L) nodes, and normal to boundary elements

Call function **discretization**:

Function `discretization`(ℓ_x,ℓ_y,IPx,IPy)**:**

- Calculate the step size in x and y-directions: $h_x = 2\ell_x/IPx$, $h_y = 2\ell_y/IPy$;
- Calculate numbers of boundary nodes $N = 2(IPx + IPy)$ and interior nodes $L = (IPx)(IPy)$
- Find the end points (x, y) of the boundary elements:
- **for** $i = 1 : IPx$ **do**
  - $x(i) = -\ell_x + (i-1)h_x$; $y(i) = -\ell_y$;
  - $x(IPx + IPy + i) = \ell - (i-1)h_x$; $y(IPx + IPy + i) = \ell_y$
- **for** $i = 1 : IPy$ **do**
  - $x(IPx + i) = \ell_x$; $y(IPx + i) = -\ell_y + (i-1)h_y$;
  - $x(2IPx + IPy + i) = \ell_x$; $y(2IPx + IPy + i) = \ell_y - (i-1)h_y$;
- Find midpoints (x_m, y_m) of boundary elements:
- $x_m(1 : N) = (x(1 : N)+x(2 : N+1))/2$; $y_m(1 : N) = (y(1 : N)+y(2 : N+1))/2$;
- Find coordinates of interior points:
- **for** $j = 1 : IPy$ **do**
  - **for** $i = 1 : IPx$ **do**
    - $x_m(N+(j-1)IPx+i) = x_m(i)$; $y_m(N+(j-1)IPx+i) = y_m(IPx+j)$;
- Find normal vector of boundary elements, $normal(nx, ny)$:
- **for** $i = 1 : N$ **do**
  - $nc(i, 1) = i$; $nc(i, 2) = i + 1$;
- **for** $i = 1 : N$ **do**
  - $N1 = nc(i, 1)$; $N2 = nc(i, 2)$; $SS = sqrt((x(N1) - x(N2))^2 + (y(N1) - y(N2))^2)$;
  - $normal(i, 1) = (y(N2) - y(N1))/SS$; $normal(i, 2) = -(x(N2) - x(N1))/SS$;
- $n_x = normal(:, 1)$; $n_y = normal(:, 2)$;
- **return** $x, y, x_m, y_m, N, L, normal, n_x, n_y$

Algorithm 3: Step 2: Construction of the discretized BEM matrices given in Eqs. (3.32) and (3.33)

Input: The discretization parameters obtained by Algorithm 2, in Step 1:
$x, y, x_m, y_m, N, L, n_x, n_y, Ha, Ha_x, Ha_y, IG$

Output: The discretized BEM matrices $\bar{H},\ G,\ H^1,\ G^1,\ HI,\ GI,\ HI^1,\ GI^1$

Call function **HG-Matrix-Generator**:

Function `HG-Matrix-Generator` $(x, y, x_m, y_m, N, L, n_x, n_y, Ha, Ha_x, Ha_y, IG)$**:**

    Construct the matrices corresponding to boundary nodes given in Eq. (3.32) :
    for $i = 1 : N$ **do**
        for $j = 1 : N$ **do**
            if $i \neq j$ **then**
                call **Function** `gauss-integral` $((x_m(i), y_m(i), x(j), y(j), x(j+1), y(j+1), n_x(j), n_y(j), Ha_x, Ha_y, Ha, IG))$**:**
                    to calculate the off-diagonal matrix entries: $\bar{H}(i, j), G(i, j), H^1(i, j), G^1(i, j)$
            else
                call **Function** `gauss-integral-diagonal` $((x(j), y(j), x(j+1), y(j+1), n_x(j), n_y(j), Ha_x, Ha_y, Ha))$**:**
                    to calculate the diagonal matrix entries: $\bar{H}(i, j), G(i, j), H^1(i, j), G^1(i, j)$

        The functions *gauss-integral* and *gauss-integral-diagonal* uses Gaussian quadrature with IG-quadrature nodes to calculate the off-diagonal and diagonal entries of the discretized BEM matrices, which are given in terms of boundary integrals in Eq. (3.31).

    Construct the matrices corresponding to interior nodes given in Eq. (3.33) :
    for $i = 1 : L$ **do**
        for $j = 1 : N$ **do**
            kk=j+1;
            call **Function** `gauss-integral` $((x_m(N+i), y_m(N+i), x(j), y(j), x(KK), y(KK), n_x(j), n_y(j), Ha_x, Ha_y, Ha, IG))$**:**
                to calculate the matrix entries: $HI(i, j), GI(i, j), HI^1(i, j), GI^1(i, j)$

    return $\bar{H},\ G,\ H^1,\ G^1,\ HI,\ GI,\ HI^1,\ GI^1$

Algorithm 4: Step 3: Construction and solution of the final linear system of equations

---

Input: All parameters and BEM matrices obtained by Algorithm 2 in Step 1 and Algorithm 3 in Step 2

Output: The approximate BEM solution for the unknowns velocity V and induced magnetic field B

Call function **solver**:

Function

`solver`$(x, y, x_m, y_m, N, L, n_x, n_y, Ha, Ha_x, Ha_y, \bar{H}, G, H^1, G^1, HI, GI, HI^1, GI^1)$:

Construct the block matrices on the left-hand side of Eq. (3.32) :

$$SYS(1:N, 1:N) \leftarrow \bar{H};\ SYS(N+1:2N, 1:N) \leftarrow G;$$
$$SYS(1:N, N+1:2N) \leftarrow G;\ SYS(N+1:2N, N+1:2N) \leftarrow \bar{H}$$

$$SYS1(1:N, 1:N) \leftarrow H^1;\ SYS1(N+1:2N, 1:N) \leftarrow G^1;$$
$$SYS1(1:N, N+1:2N) \leftarrow G^1;\ SYS1(N+1:2N, N+1:2N) \leftarrow H^1$$

Insert the boundary conditions $V = 0$ and $B = 0$ to the system, and shuffle the columns and rows of the matrices SYS and $SYS1$ corresponding to known entries of vectors $U = [V \ \ B]'$ and its normal derivative Q. Then, the final system is constructed as:

$$RHS_{vector} = SYS1 \times Q;\ LHS_{system} = SYS;$$

Solve the system by using MATLAB built-in function **mldivide** to obtain the solution on the boundary, where $solution_{boundary} = [U_b \ \ Q_b]'$:

$$solution_{boundary} = LHS_{system} \backslash RHS_{vector}\ ;$$

Use boundary values found above to obtain the values of V and B at interior nodes by inserting them in Eq. (3.33) :

Construct the block matrices on the right-hand side of Eq. (3.33) :

$$SYSA(1:N, 1:N) \leftarrow HI;\ SYSA(N+1:2N, 1:N) \leftarrow GI;$$
$$SYSA(1:N, N+1:2N) \leftarrow GI;\ SYSA(N+1:2N, N+1:2N) \leftarrow HI$$

$$SYSA1(1:N, 1:N) \leftarrow HI^1;\ SYSA1(N+1:2N, 1:N) \leftarrow GI^1;$$
$$SYSA1(1:N, N+1:2N) \leftarrow GI^1;\ SYSA1(N+1:2N, N+1:2N) \leftarrow HI^1$$

Calculate the solution at interior nodes, $solution_{interior} = [U_i \ \ Q_i]$ by using Eq. (3.33):

$$solution_{interior} = SYSA \times U_b + SYSA1 \times Q_b;$$

Assign the obtained boundary $[U_b \ \ Q_b]$ and interior solution $[U_i \ \ Q_i]$ to V and B:

$$V(1:N) = U_b(1:N);\ B(1:N) = U_b(N+1:2N);$$
$$V(N+1:N+L) = U_i(1:L);\ B(N+1:N+L) = U_i(L+1:2*L)$$

return *V and B*

Chapter 4
MHD Channel Flows

In this chapter, steady forms of the MHD duct flow equations (1.57) given in Sect. 1.2.8 are solved using BEM with the fundamental solution of coupled MHD equations derived in Chap. 3. Pressure driven or electrically driven MHD flows in rectangular ducts or in infinite regions are all considered, and numerical solutions are obtained for increasing values of Hartmann number. Special emphasis is given to partly insulated and partly perfectly conducting walls. Convergence of infinite integrals is shown, and the thickness of parabolic boundary layers is computed. MHD channel flows are also studied by some other numerical methods. The FEM solutions are given in [46] and [47] for moderate values of Hartmann number, and the stabilized FEM with residual-free bubble functions increases Hartmann number up to 1000 in [48]. Upwinding meshfree point collocation method and radial point interpolation methods [49, 50] are used for solving fully developed steady MHD flow in rectangular ducts at high Hartmann numbers.

We consider first a fully developed flow in a long channel with rectangular or circular cross section (duct). An electrically conducting fluid is driven through the duct by a constant pressure gradient or electrodes placed in the walls of the channel. The fluid is exposed to an externally applied magnetic field, and it flows in the direction of the channel axis (z-axis), so its space variation is only in the duct region. Duct walls on which the applied magnetic field is perpendicular are the Hartmann walls, whereas the walls parallel to the external magnetic field are the side walls. Thus, the fluid has an unidirectional velocity $\mathbf{u} = (0, 0, V(x, y))$, and the magnetic field $\mathbf{B} = (B_x, B_y, B(x, y))$ has unknown component $B(x, y)$ only in the z-direction with known components $B_x = B_0 \cos\gamma$, $B_y = B_0 \sin\gamma$. B_0 is the intensity of the applied magnetic field, and γ is the angle made with the direction of B_0 and the y-axis.

The magnetic field and moving fluid's interaction produce an electric field $\mathbf{u} \times \mathbf{B}$ that drives the electric current $\mathbf{J}$. A Lorentz force $\mathbf{J} \times \mathbf{B}$ is generated by current components that are perpendicular to the magnetic field lines. In the center of the duct, the Lorentz force acts in the direction opposite to the flow direction slowing

M. Tezer-Sezgin, C. Bozkaya, *Boundary Element Method for Magnetohydrodynamic Flow*, Surveys and Tutorials in the Applied Mathematical Sciences 14,
https://doi.org/10.1007/978-3-031-58353-7_4

down the flow, while the main balance of forces is established between the Lorentz force and the driving pressure gradient. To meet the kinematic no-slip boundary conditions, the velocity rapidly decreases within thin boundary layers near the Hartmann walls. However, there are cases where the velocity slips in the adjacent walls according to the material of the walls made of. The electric conductivity of channel walls affects how current is distributed in the fluid and establishes the flow pattern. For insulating walls, the current closes through the relatively thin Hartmann and side layers. Since these layers are so thin, they have a high electric resistance, which reduces the current's magnitude. For perfectly conducting walls, the currents close through the walls. This increases the magnitude of currents considerably. For variably conducting walls, a significant portion of current may close through the walls, increasing again the total magnitude of currents compared to insulating conditions. Thus, stronger Lorentz forces are expected with increasing wall conductance. For the slipping and the variably conducting duct walls, the corresponding wall conditions are $\alpha \partial V/\partial n + V = 0$ and $\partial B/\partial n + cB = 0$, respectively, with slip length α and conductivity constant parameter c. We consider all the possible cases of wall slip and wall conductivity by taking several values of α and c in the MHD rectangular duct flow problems. It is noted that, except for simple boundary conditions of the duct (for example, no-slip and insulated walls), there is no analytical solution for the MHD duct flow problems. Thus, for the most general form of slip and conductivity of the duct walls, some numerical methods must be employed for the solution of the MHD flows.

When the Hartmann walls are extended to infinity, the MHD flow is turned out to be the MHD flow between parallel plates, and when one of these parallel plates is also extended vertically to infinity, it is turned out to be the MHD flow in an infinite region over a plate. In these problems, the behaviors of V and B as $x \to \pm\infty$ and also as $y \to \infty$ are examined in terms of the convergence of the integrals on the infinite Hartmann walls and the convergence of infinite boundary integrals on the upper half plane using a fictitious boundary, respectively. The cases of partly insulated–partly perfectly conducting plates are also considered to see the effect of these wall conditions on the solution. Thickness of boundary layers, emanating from the points where the conductivity changes, is also computed from the solution of BEM discretized matrix–vector systems. In all the considered MHD problems, the effect of Hartmann number is analyzed and discussed, and the orientation of the applied magnetic field is included to the discussions.

4.1 Pressure Driven MHD Flow in Channels

A straight channel of rectangular or circular cross section (duct) with thin walls of constant thickness is considered. The channel is of sufficient length that the end effects are neglected. The external magnetic field of strength B_0 is supposed to be transverse and constant along the channel (perpendicular to the channel axis). We assume that the fluid motion is due to a constant pressure gradient, and the flow is fully developed through the channel, i.e., $\partial/\partial z = 0$. The problem is governed by the steady form of the MHD flow equations that are given in Eq. (1.57) as

$$\left.\begin{aligned}\nabla^2 V + Ha_x\frac{\partial B}{\partial x} + Ha_y\frac{\partial B}{\partial y} &= -1\\ \nabla^2 B + Ha_x\frac{\partial V}{\partial x} + Ha_y\frac{\partial V}{\partial y} &= 0\end{aligned}\right\} \quad \text{in } \Omega\,. \tag{4.1}$$

4.1.1 MHD Flow in Rectangular Duct

A square duct $\Omega = \{(x, y) \mid -1 < x, y < 1\}$ of side length two, displayed in Fig. 4.1, is taken in the xy-plane with the channel axis perpendicular to the duct at the center. External magnetic field applies with an angle γ made with the y-axis. No-slip velocity condition ($V = 0$ on Γ) is usually taken for the velocity or partly no-slip and partly slip velocity conditions are imposed. All kinds of conductivities as insulated ($B = 0$), perfectly conducting ($\frac{\partial B}{\partial n} = 0$), and variably conducting ($\partial B/\partial n + cB = 0$) are considered at the duct walls.

The validation of the numerical algorithm is assessed by means of the MHD flow problem in a duct with no-slip and electrically insulated walls ($\sigma_{\text{wall}} = 0$, that is $c \to \infty$) in the presence of a horizontally applied external magnetic field ($\gamma = \pi/2$), which is considered as the benchmark MHD duct flow problem. A theoretical solution for this case has been given by Shercliff [51] in terms of Fourier series expansions. The present numerical solution is based on the BEM with the fundamental solution (3.24) of coupled steady MHD equations (4.1) [45]. A quantitative comparison of the numerical algorithm for BEM is performed using the relative errors between the exact and approximate BEM values of V and B,

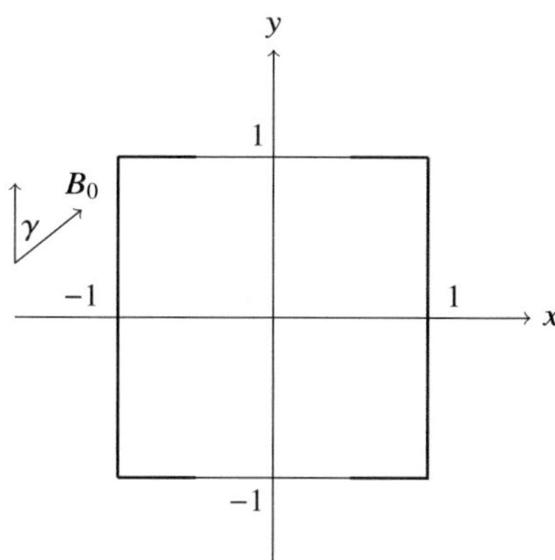

Fig. 4.1 Geometry of MHD flow in rectangular duct

Table 4.1 Comparison of exact and BEM solutions in terms of relative errors and volumetric flow rates at $Ha = 10, 50, 100, 300$

Ha	$Rele_V$	$Rele_B$	$Q^{\text{analytical}}$	Q^{BEM}
10	0.0054449	4.3367×10^{-6}	0.2606	0.2606
50	0.014595	4.1726×10^{-5}	0.068881	0.06875
100	0.039999	8.7218×10^{-5}	0.036509	0.036118
300	0.31199	0.00021708	0.014172	0.012598

$$Rele_V = \max_{P\in\Omega}\left|\frac{V^{\text{exact}}(P) - V^{\text{BEM}}(P)}{V^{\text{exact}}(P)}\right|, \qquad Rele_B = \max_{P\in\Omega}\left|\frac{B^{\text{exact}}(P) - B^{\text{BEM}}(P)}{B^{\text{exact}}(P)}\right|$$

and the volumetric flow rate $Q = \int_\Omega V\,d\Omega$. Table 4.1 shows the relative errors $Rele_V$ and $Rele_B$ as well as the dimensionless volumetric flow rate Q at various Hartmann numbers. The relative error in the velocity is around 10^{-2}, while it is around 10^{-5} in the induced magnetic field when an adequate fixed number of boundary elements, namely $N = 324$, are employed in the calculations. The well agreement of the present numerical findings is further observed in terms of volumetric flow rates obtained by BEM, Q^{BEM}, with the analytical ones, $Q^{\text{analytical}}$, computed from the theoretical series solution presented in Shercliff [51]. It is also noted that the flow rate reduces as the intensity of the magnetic field increases.

On the other hand, the number of boundary elements N at $Ha = 100$ and $\gamma = \pi/2$ is chosen based on the tests on the distributions of mean value of velocity, V_{mean}, and absolute mean value of induced magnetic field, $|B|_{\text{mean}}$, with respect to the number of boundary elements N. That is, Fig. 4.2 shows the variations of the V_{mean} and $|B|_{mean}$ from the exact and BEM solutions with respect to N, while Fig. 4.3 shows the variations of relative errors in V and B versus N. It is observed that when $N \approx 324$, sufficient convergence to the precise values is reached, and larger values of N no longer produce a noticeable change in the mean values of the velocity and the induced magnetic field. Thus, in the subsequent computations, the boundary of the domain is discretized by using $N \leq 324$ boundary elements at each $Ha \leq 100$. Additionally, when the boundary is discretized with an appropriate number of elements, Fig. 4.3 shows that the BEM solutions agree well with the exact values once again.

First, the influence of the wall conductivity on the flow and the induced magnetic field for the MHD flow is considered at various Ha. In this respect, Fig. 4.4 shows the equi-velocity lines and the induced magnetic field contours for increasing values of Ha, for both insulating ($B = 0$) and perfectly conducting ($\partial B/\partial n = 0$) walls. Due to the simplifications and the geometry of the problem, the contours of the scalar B coincide with the field lines of the current density and referred as induced current. For the insulating walls, since no currents can enter the walls, the currents in the duct are closed in front of the Hartmann walls. There, the Lorentz and pressure forces are balanced by the friction force resulting in thin velocity boundary layers with steep velocity gradients. Velocity is high at the core of the duct and reduces to zero at the walls. Induced magnetic field lines are nearly aligned with the applied magnetic field in front of the side walls. As Ha increases, the magnitudes of both

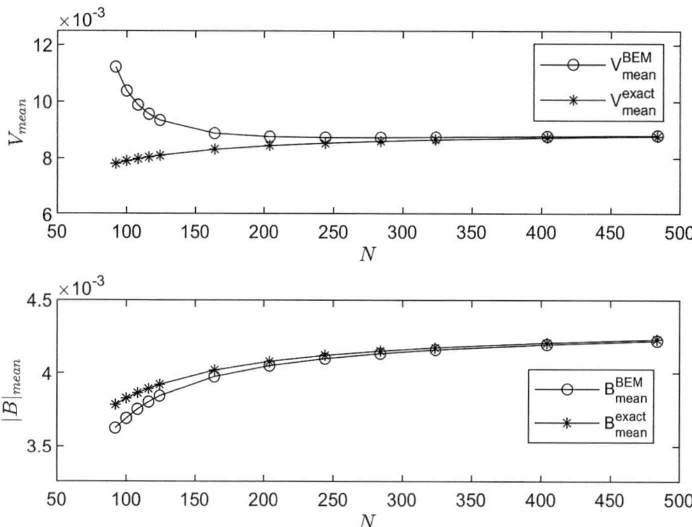

Fig. 4.2 Grid independence test in terms of V_{mean} and B_{mean} when $Ha = 100$, $\gamma = \pi/2$

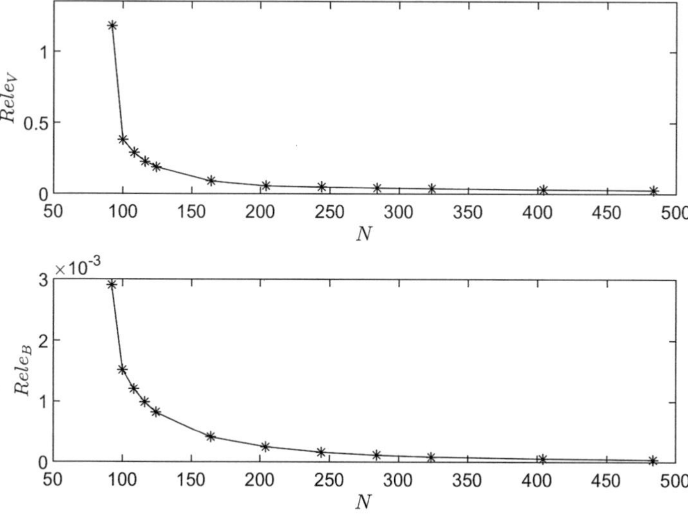

Fig. 4.3 Grid independence test in terms of $Rele_V$ and $Rele_B$ when $Ha = 100$, $\gamma = \pi/2$

velocity and induced magnetic field drop indicating a flattening flow. The boundary layers are more pronounced when Ha is increasing leaving the core of the duct almost stagnant. The thickness of Hartmann layers is of order $\frac{1}{Ha}$, whereas the thickness of side layers is of order $\frac{1}{\sqrt{Ha}}$ [51]. Figure 4.4(b) demonstrates the flow and induced current behaviors for perfectly conducting ($\sigma_{\mathrm{wall}} \to \infty$) duct walls.

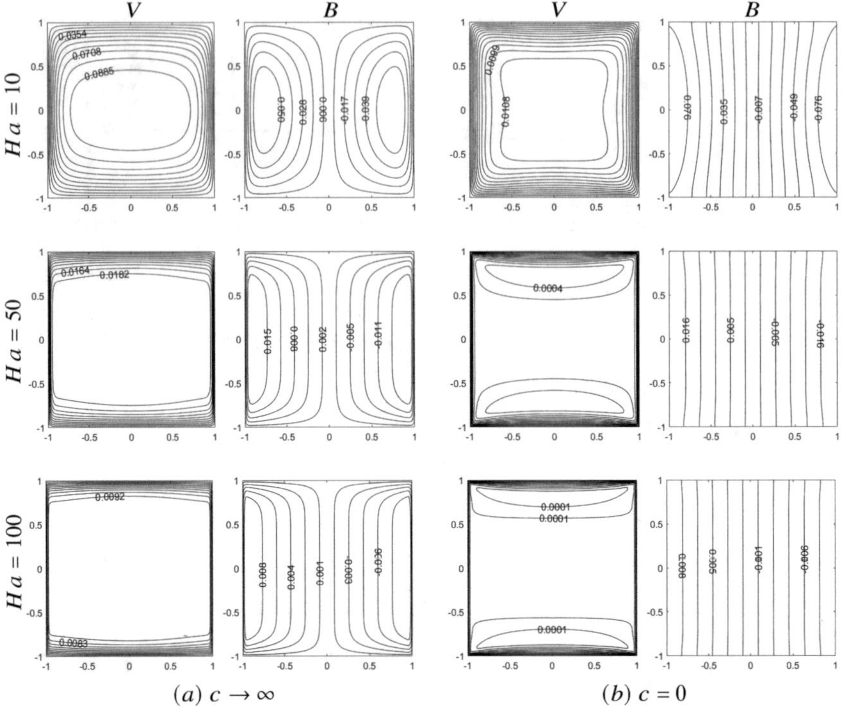

Fig. 4.4 Equi-velocity and current lines at $Ha = 10, 50, 100$, $\gamma = \pi/2$: (**a**) insulating walls ($c \to \infty$) and (**b**) perfectly conducting walls ($c = 0$)

This case has been considered by Hunt [52] and Chang and Lundgren [53]. The BEM solution computes B values on the walls since $\frac{\partial B}{\partial n}$ is assigned. Tezer-Sezgin and Dost [54] have given a BEM solution for arbitrary wall conductivity case by decoupling the MHD flow equations, but then boundary conditions are coupled and fundamental solution of modified Helmholtz equation has been used. Here, we demonstrate the BEM solution that is obtained by using the fundamental solution of MHD equations in coupled form. The flow is mostly concentrated in front of the side walls in terms of two loops obeying again boundary layer thicknesses $\frac{1}{Ha}$ and $\frac{1}{\sqrt{Ha}}$ for Hartmann and side layers, respectively. Again velocity and induced magnetic field decrease as Ha increases. Induced currents emanate from one side wall and enter to the other side wall. As Ha increases, they are almost perpendicular to the applied field.

In Fig. 4.5, the effect of the direction of the external magnetic field is shown on the flow and induced magnetic field for $Ha = 50$ and for insulating duct walls. It is noticed that the flow and the induced current align in the direction of applied field. As the angle made with the y-axis decreases, Hartmann layers are pushed through the left bottom and right upper corners and finally settle at the bottom and top walls for $\gamma = 0$. Side layers are getting thicker in front of the left upper and right lower

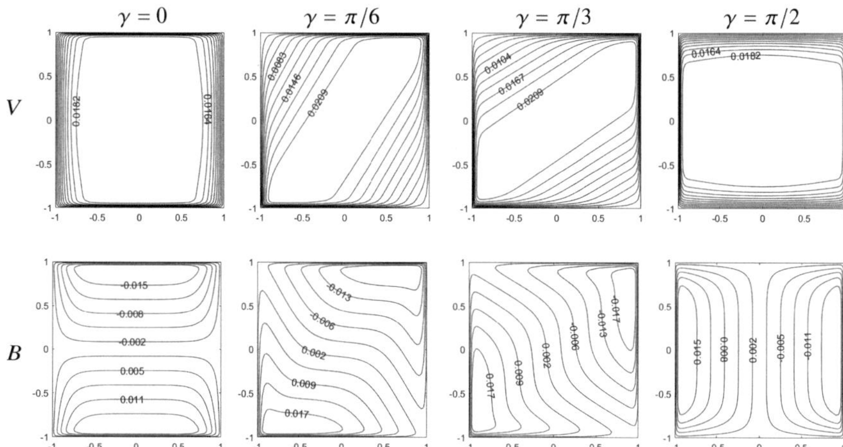

Fig. 4.5 Equi-velocity (top) and current lines (bottom) at $Ha = 50$ with insulating walls ($c \to \infty$), $\gamma = 0, \pi/6, \pi/3, \pi/2$ (from left to right)

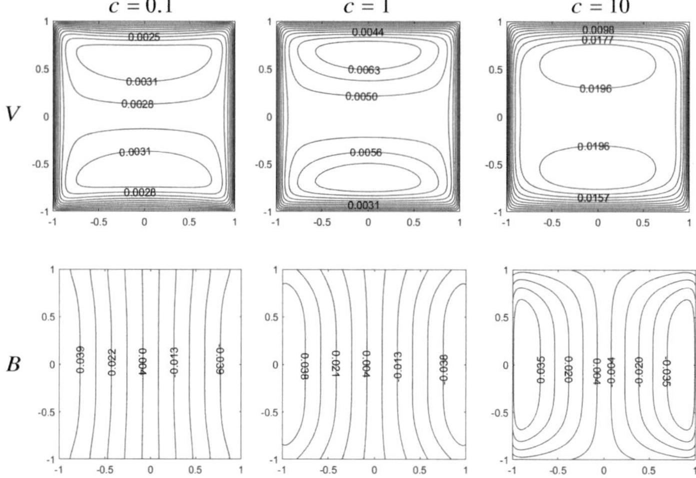

Fig. 4.6 Equi-velocity (top) and current lines (bottom) at $Ha = 20$, $\gamma = \pi/2$, for variably conducting walls ($c = 0.1, 1, 10$, from left to right)

corners and settle at the right and left walls for $\gamma = 0$. The effect on the current lines is similar, that is, they are distributed again from left bottom corner to right upper corner, finally close in front of top and bottom walls symmetrically for $\gamma = 0$.

Figure 4.6 demonstrates the equal velocity and magnetic induction lines for variably conducting duct walls when $Ha = 20$. One can see that fluid velocity decreases with increasing wall conductivity (i.e., $c \to 0$). However, the magnitude of current induction stays almost the same, but its profile significantly changes from

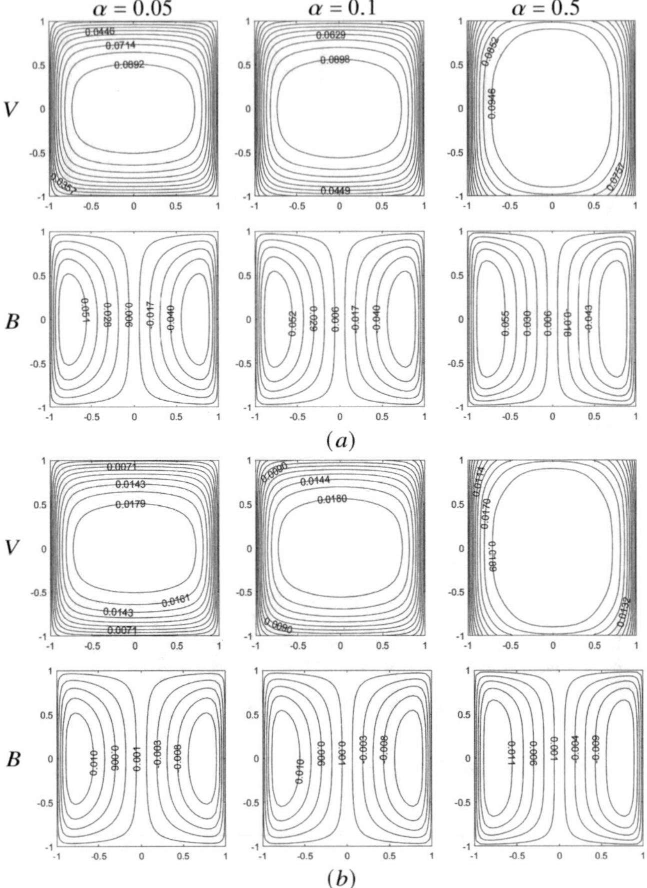

Fig. 4.7 Equi-velocity and current lines at (**a**) $Ha = 10$ and (**b**) $Ha = 50$, $\gamma = \pi/2$ with insulating walls ($c \to \infty$) and slip length $\alpha = 0.05, 0.1, 0.5$ (from left to right)

being two closing loops in front of the Hartmann walls to straight lines emanating from one side wall and entering to the other side wall. These behaviors are in agreement with the ones obtained by BEM solution of decoupled MHD modified Helmholtz equations [54].

In Fig. 4.7 we show behaviors of the flow and the induced magnetic field when the velocity slips near the side walls. It is observed that for a fixed Hartmann number, the increase in slip length α causes an increase in the velocity magnitude. Also, stagnant flow region at the center is enlarged and elongated with an increase in the slip length. However, for larger Hartmann number values, this increase in the magnitude of velocity is lessened. Although the magnitude of the induced magnetic field is not affected much, the current lines are elongated through the corners with an increase in the slip length.

4.1.2 MHD Flow in Circular Duct

In this problem, the channel is a long pipe of circular cross section (circular duct), and the MHD flow equations (4.1) are solved using BEM with the fundamental solution of coupled MHD equations without the need of decoupling the equations. The boundary conditions are taken as $V = 0$, $B = 0$ on Γ implying no-slip velocity and insulated wall case. Gold [55] proposed an exact solution for this case of MHD flow that was obtained using a Fourier analysis.

Figure 4.8 depicts the velocity and induced current behaviors for increasing values of Ha. The fluid becomes stagnant, and the velocity takes its maximum value at the center of the circular duct. Boundary layers are developed for both the velocity and the induced magnetic field as Ha increases. The induced current closes itself inside the duct in terms of two loops in the direction of the applied magnetic field. The velocity action is completely near the top and bottom parts of the circular duct in terms of boundary layers. The flattening tendency of the flow is also observed when Ha increases [56, 57].

In Fig. 4.9 equi-velocity and current lines for $Ha = 50$ are presented for $c \to \infty$ (insulated) wall through $c = 0$ (perfectly conducting) wall. The boundary layer formation for the flow is more pronounced when the wall conductivity is increasing ($c \to 0$). Current lines are deformed from the two loops profile and are emanating from the duct straightly to the exterior when the conductivity of the duct wall increases. The flow is mostly stagnant in the duct when the wall conductivity is increasing.

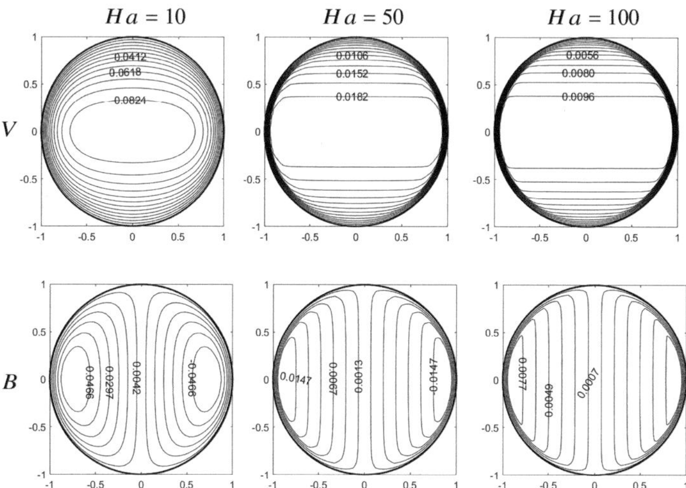

Fig. 4.8 Equi-velocity and current lines for $Ha = 10, 50, 100$, $\gamma = \pi/2$, $c \to \infty$ for MHD flow in circular duct

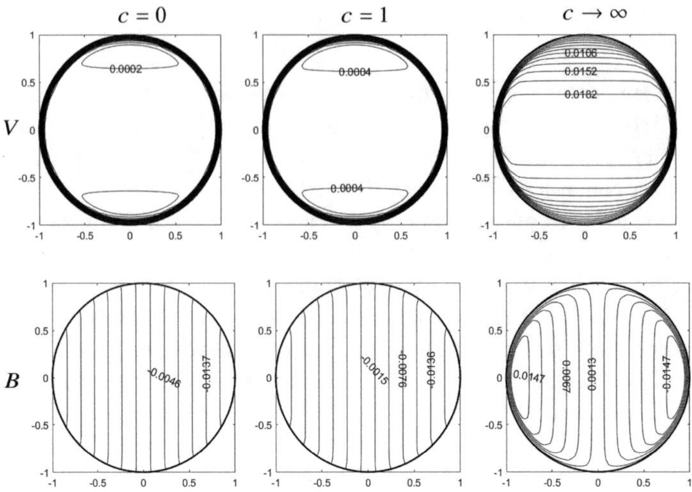

Fig. 4.9 Equi-velocity and current lines for $Ha = 50$, $\gamma = \pi/2$, $c = 0$, $c = 1$, $c \to \infty$ for MHD flow in circular duct

4.1.3 MHD Flow in Rectangular Duct with Perfectly Conducting Side Walls and Insulating Hartmann Walls

It is assumed that the side walls parallel to external magnetic field are perfectly conducting and Hartmann walls are insulating for which the coupled MHD equations (4.1) cannot be decoupled. An integral form of solution is established by Grinberg [58] with a Green's function in terms of double infinite series. Its practical usefulness is limited and cannot be solved easily as stated by Dragos [7] since it is a rigorous solution. This MHD flow problem has been solved in the work of Sezgin [59] by reducing the equations to decoupled modified Helmholtz equation but with coupled boundary conditions. In that study, the BEM solution could have been obtained only for the values of $Ha \leq 10$. The BEM with a fundamental solution for coupled MHD equations can easily treat the whole equations with these boundary conditions [60].

Figure 4.10 shows the flow and magnetic induction profiles for $Ha = 50$ and $Ha = 100$. The flow behavior is the same as in the case when all the walls are insulating. Thus, the increase in Ha reduces the velocity magnitude flattening the flow. Hartmann and side layers are formed when Ha increases. Induced current lines deviate from being two loops symmetrically concentrated in front of the Hartmann walls to straight lines at the center of the duct entering from one side wall through the other since current can enter from these walls. Again Lorentz and pressure forces are balanced near the Hartmann walls.

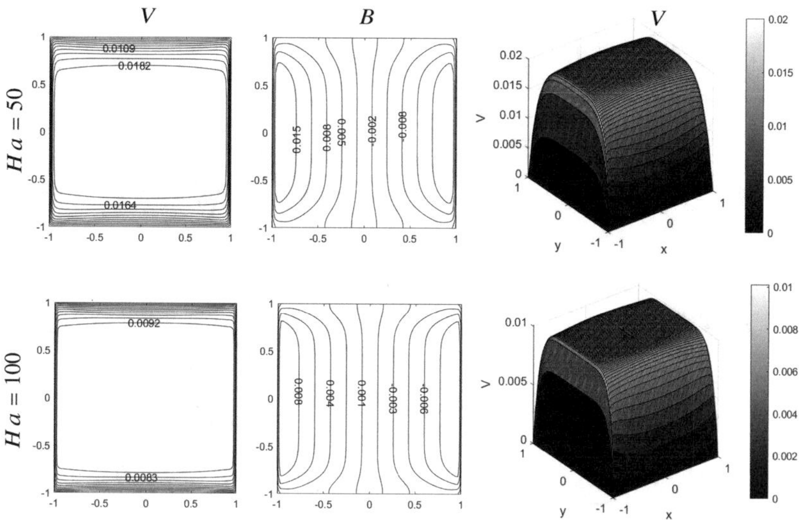

Fig. 4.10 Equi-velocity and current lines for $Ha = 50$ (top) and $Ha = 100$ (bottom), $\gamma = \pi/2$, for MHD flow in duct with perfectly conducting side and insulating Hartmann walls

4.1.4 MHD Flow in Rectangular Duct with Insulating Side Walls and Perfectly Conducting Hartmann Walls

In this case, the duct is considered with perfectly conducting Hartmann walls ($\sigma_{\text{wall}} \to \infty$, $c = 0$) and insulating side walls ($\sigma_{\text{wall}} = 0$, $c \to \infty$). Again, the coupled MHD equations present difficulties in solving with these boundary conditions. Hunt [52] gave a solution for this configuration and showed that large positive and negative velocities are induced at high values of Hartmann number. We obtain the BEM solution with the fundamental solution of coupled MHD equations. Magnetic induction lines displayed in Fig. 4.11 close through the Hartmann walls and form side layers near the bottom and top walls. Flow concentrates symmetrically in front of the side walls in terms of two loops. At the central part of the duct, the pressure gradient is balanced by the Lorentz force. Near the side walls Lorentz forces are reduced resulting in higher velocities with the side jets. The velocity level curves along the x-axis resemble the letter M, and the flow profile is called the M-shape profile.

Some common behavior are observed from the MHD flow in rectangular ducts that are presented in Sects. 4.1.1 and 4.1.3–4.1.4 (insulating, perfectly conducting or two opposite walls are insulating the other two perfectly conducting). These are, for moderate to high Hartmann numbers:

- The velocity distribution is flat at the center of the duct, that is, the fluid is stagnant.

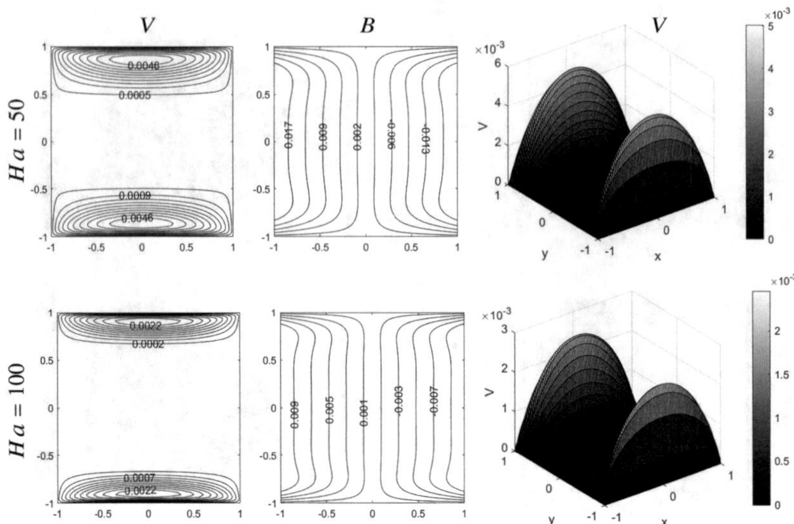

Fig. 4.11 Equi-velocity and current lines for $Ha = 50$ (top) and $Ha = 100$ (bottom), $\gamma = \pi/2$, for MHD flow in duct with insulating side and perfectly conducting Hartmann walls

- As Ha increases, boundary layers are formed. The thickness of boundary layers decreases with increasing intensity of applied magnetic field.
- The current density is nearly constant in most of the duct area.

4.1.5 MHD Flow in Rectangular Duct with Partly Insulated–Partly Perfectly Conducting Hartmann Wall

The boundary element method has been applied with the fundamental solution of coupled steady MHD equations (4.1) that was established by Bozkaya and Tezer [45]. That is, it is possible to solve MHD duct flow problems with the most general form of wall conductivities and for large values of Hartmann number. For this partly insulated–partly conducting one side of the duct, the MHD flow problem does not have an exact solution. The solution for this MHD problem has been obtained in the work of Sezgin [61] by reducing the equations to dual series equations on this boundary and then to a Fredholm integral equation of the second kind that was solved numerically.

The MHD flow in a rectangular channel with a cross section $0 \leq x \leq 1$, $-0.5 \leq y \leq 0.5$, that is, $a = 1$, $b = 0.5$, subjected to a magnetic field in the x-direction (i.e., $\gamma = \frac{\pi}{2}$) is displayed in Fig. 4.12. The no-slip condition $V = 0$ is imposed everywhere on the walls that are insulated ($B = 0$) except the portion of length ℓ on $x = 0$ line, symmetrically about the origin (i.e., $\frac{\partial B}{\partial n} = 0$ on $x = 0$, $-\ell \leq y \leq \ell$).

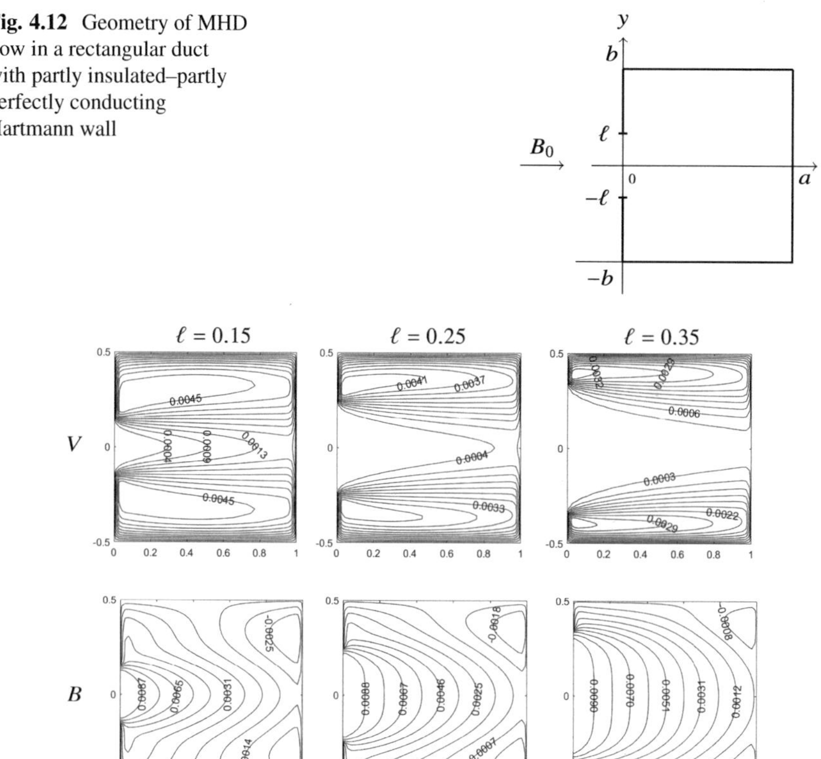

Fig. 4.12 Geometry of MHD flow in a rectangular duct with partly insulated–partly perfectly conducting Hartmann wall

Fig. 4.13 Equi-velocity (top) and current lines (bottom) for $Ha = 100$, $\gamma = \pi/2$, $\ell = 0.15, 0.25, 0.35$ (from left to right)

In Fig. 4.13 the velocity and induced magnetic field contours are presented for $Ha = 100$ and $\ell = 0.15, 0.25, 0.35$. The enlargement of the stagnant region for the velocity in front of the conducting portion is well observed as the length of it increases. Further, current lines are pushed through the center of the duct with the effect of conducting portion on the left Hartmann wall. As ℓ is increased, they cover almost all parts of the duct squeezing the right current loop near the insulating wall $x = 1$ to two small vortices at the right corners. The Lorentz and pressure forces are balanced again at the center of the duct between the side layers as the conducting portion enlarges. There are parabolic boundary layers emanating from the discontinuity points $y = -\ell$ and $y = \ell$ in the direction of the applied magnetic field again for both the velocity and the induced magnetic field. The thickness of these parabolic boundary layers is going to be computed in terms of Hartmann number in Sect. 4.2.4 for MHD flow on the upper half plane.

4.1.6 MHD Flow in Electrodynamically Coupled Rectangular Ducts

The problem considered here is the electrodynamic coupling of MHD flows of electrically conducting fluids in coupled rectangular ducts separated by a barrier that is partially perfect conductor and partially insulator as visualized in Fig. 4.14. The two flows are separated from each other by this thin barrier under the effect of a horizontally applied external uniform magnetic field of constant strength B_0. The fluids in duct I and duct II are driven by constant pressure gradients that are produced by a pump and the hydraulic load of duct II, respectively. The transfer of hydraulic energy to channel II is done electrodynamically since part of the current induced by flow I, through the conducting partition ($|y| < \ell$, $x = 0$), forms a closed current loop that passes into channel II, and which by interacting with the transverse magnetic field generates a force in the direction of flow I (positive z-axis). Butsenicks and Shcherbinin [62] solved the problem for the case of the barrier separating the flows being completely perfect conductor. They have encountered great difficulties due to the alternating asymptotic series in the solution for large Ha. Sezgin [63] solved the equations by reducing them into a set of dual integral equations and then to a Fredholm integral equation that is solved numerically. On the other hand, we have solved the problem for a barrier that can be partly perfect conductor and partly insulator by using the BEM with the fundamental solution of coupled MHD equations with details given in the work of Bozkaya and Tezer-Sezgin [64].

The nondimensional system of governing equations defining the uniform flows in channels I and II are given as [7]

$$\left.\begin{aligned} \nabla^2 V + Ha\frac{\partial B}{\partial x} &= -1 \\ \nabla^2 B + Ha\frac{\partial V}{\partial x} &= 0 \end{aligned}\right\} \quad \text{in } \Omega_1 \text{ (Channel I)} \tag{4.2}$$

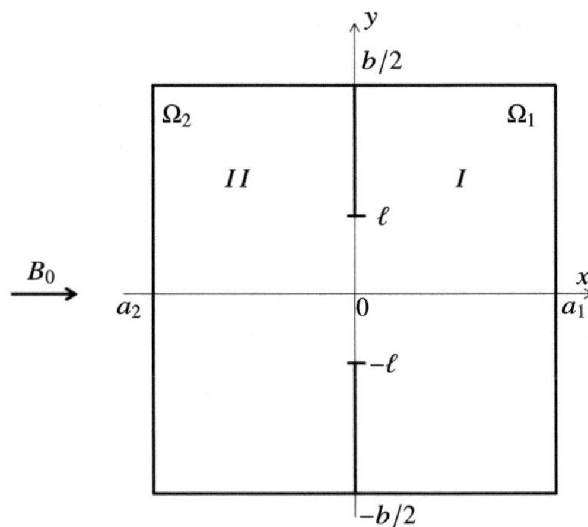

Fig. 4.14 Problem geometry for the MHD flow in electrodynamically coupled rectangular ducts

$$\left.\begin{aligned}\nabla^2\bar{V} + Ha\frac{\partial \bar{B}}{\partial x} &= s \\ \nabla^2\bar{B} + Ha\frac{\partial \bar{V}}{\partial x} &= 0\end{aligned}\right\} \quad \text{in } \Omega_2 \text{ (Channel II)}, \tag{4.3}$$

where V, $\bar{V}$ and B, $\bar{B}$ are the z-components of the velocity and magnetic field vectors in Ω_1, Ω_2, respectively. The ratio of pressure gradients is $s = -\frac{\partial p_2}{\partial z} / \frac{\partial p_1}{\partial z}$. The parameter s is actually positive, since the pressure gradient produced in channel II is in the direction of the fluid motion, while in channel I the direction of the pressure gradient is opposite to the flow direction. The boundary conditions for the systems (4.2)–(4.3) are the following: The velocity everywhere at the solid walls including the barrier is zero; the induced magnetic field is zero everywhere at the insulating walls and continuous on the conducting partition separating the flows, that is,

$$\begin{aligned} B(0, y) &= \bar{B}(0, y) & -\ell \le y < \ell \\ \frac{\partial B}{\partial n}(0, y) &= \frac{\partial \bar{B}}{\partial n}(0, y) & -l \le y < l\,. \end{aligned} \tag{4.4}$$

The computational domain is determined by taking $b = a_1 = 1$, $a_2 = -1$ in the numerical simulations by BEM.

Equal velocity and magnetic field lines have been drawn in Fig. 4.15 for $Ha = 10, 50, 100$ under horizontally applied magnetic field when the pressure ratio of the two ducts is $s = 0.5$, and the length of the perfect conductor partition is $\ell = 0.15$. An increase in Ha leads to the formation of boundary layers near the boundaries and parabolic layers in front of the discontinuous conductivity points as in the case of one rectangular duct problem given in Sect. 4.1.5. However, because $s < 1$, the electromagnetic effects are more dominant in the duct II. The equal velocity lines are visibly affected by the magnetic field in duct II as compared with duct I for $Ha = 10$. For larger values of Ha, electromagnetic effects predominate the effects of pressure gradient in both the ducts. Moreover, an increase in Ha gives rise to reversal of flow direction in duct II, in region adjoining the conducting part. The effect of varying Ha on current lines (equal magnetic field lines) is the same as the case of one rectangular duct. The current induced in duct I is partially closed (negative lines), and the rest of the current is connected to the fluid in duct II through the perfectly conducting partition.

Figure 4.16 presents the pattern of velocity and induced magnetic field for an increased length of perfect conductor on the separating barrier with $\ell = 0.35$ for increasing Ha values and fixed $s = 0.5$. The velocity behaviors in both of the ducts are similar to that observed in one rectangular duct. That is, as Ha increases, the parabolic boundary layers are more pronounced. The central parts of the two ducts show that the fluids are almost stagnant in these areas that enlarge as the length of

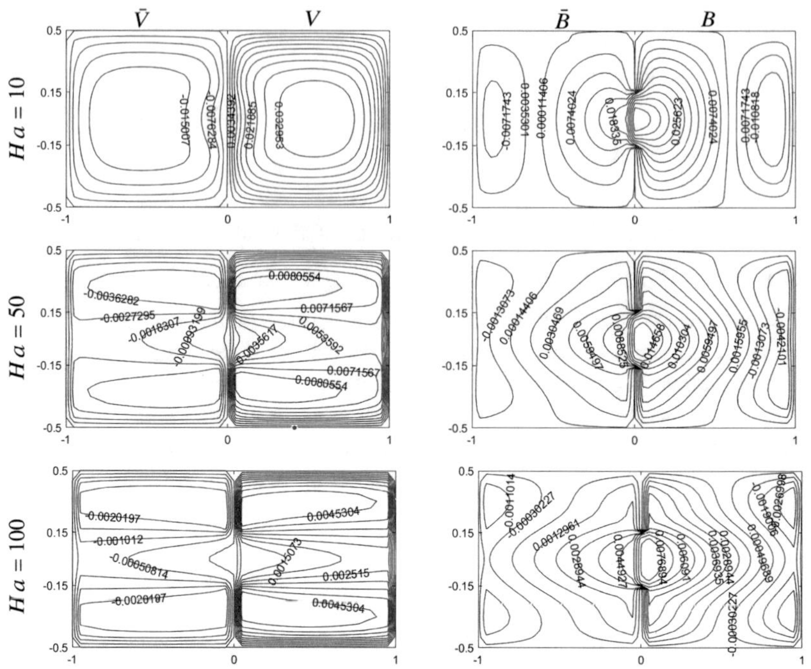

Fig. 4.15 Equi-velocity (left) and current lines (right) for $Ha = 10, 50, 100$, $\gamma = \pi/2$, $\ell = 0.15$, $s = 0.5$

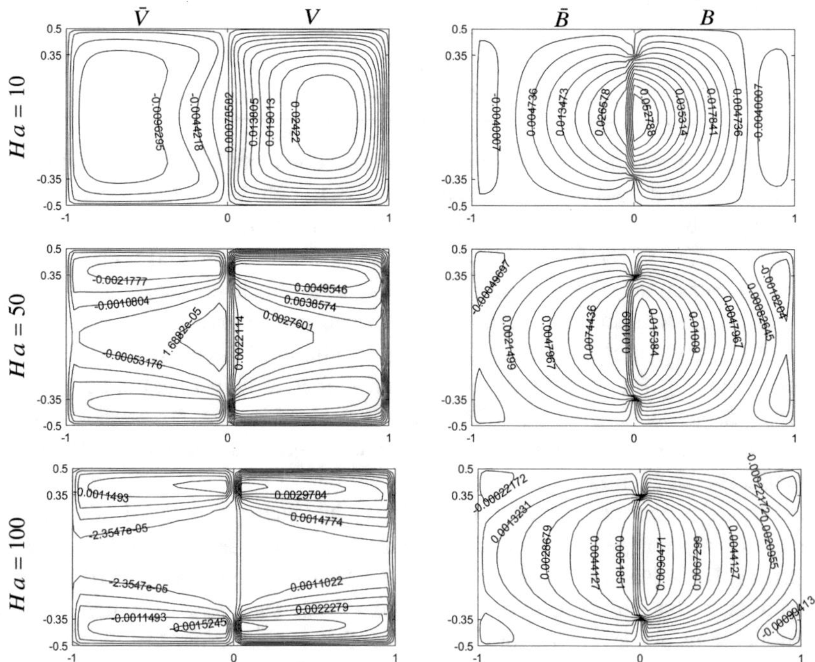

Fig. 4.16 Equi-velocity (left) and current lines (right) for $Ha = 10, 50, 100$, $\gamma = \pi/2$, $\ell = 0.35$, $s = 0.5$

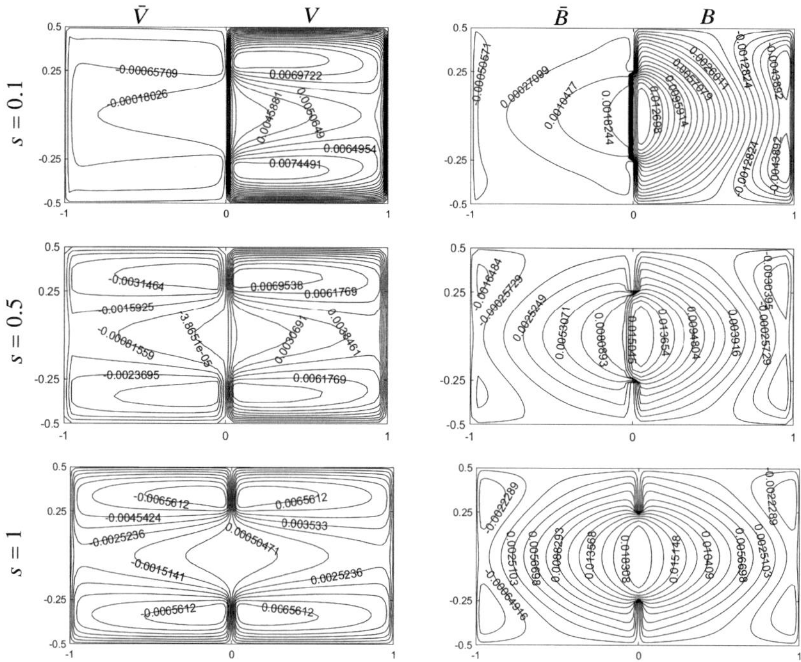

Fig. 4.17 Equi-velocity (left) and current lines (right) for $Ha = 50$, $\gamma = \pi/2$, $\ell = 0.25$, $s = 0.1, 0.5, 1$

the conducting portion increases. Current lines of both ducts again connect through the conducting portion of the barrier.

Figure 4.17 shows velocity and induced current behaviors for several values of s that are 0.1, 0.5, and 1 for a fixed $Ha = 50$ and $\ell = 0.25$. When $s = 1$, the negative velocities in duct II are equal in magnitude to the positive velocities in duct I. The maximum contrast occurs when $s = 1$ and $s = 0.1$. This means that when the pressure in the second duct is reduced by a ratio of $1 : 10$, there is almost complete reversal of flow in duct II and the flow in duct I being affected only marginally. However, in duct II the flow in the positive direction is confined to a very narrow region near the insulated parts of the partition. It is also noted that in duct II, the velocity is lower in the center. Current lines for $s = 0.1$ do not form vortices with negative B values. Maximum B occurs in front of the conducting portion in duct I.

The integrated flow rates in both ducts calculated from

$$Q_1 = \int_{-1/2}^{1/2} \int_0^1 V(x, y)\, dx dy, \qquad Q_2 = \int_{-1/2}^{1/2} \int_{-1}^0 \bar{V}(x, y)\, dx dy$$

are tabulated in Table 4.2 for the values of $10 \leq Ha \leq 100$, $0.15 \leq \ell \leq 0.35$, and $0.1 \leq s \leq 0.9$. It is noticed from Table 4.2 that for a fixed Ha and ℓ, the increase in s decreases the flow rates Q_1 and Q_2 in both of the ducts. That is, the pressure

Table 4.2 Integrated flow rates in coupled ducts

Ha	ℓ	s	Q_1	Q_2
50	0.25	0.1	0.0056	−0.0005
		0.3	0.0052	−0.0013
		0.5	0.0048	−0.0020
		0.8	0.0043	−0.0032
		0.9	0.0041	−0.0036
50	0.15	0.5	0.0062	−0.0029
	0.25		0.0048	−0.0020
	0.35		0.0037	−0.0013
10	0.25	0.5	0.0181	−0.0095
20			0.0106	−0.0049
50			0.0048	−0.0020
100			0.0026	−0.0011

gradient in duct I is higher than the one in duct II. When the conducting portion of the barrier is enlarged, or Ha increases, the flow rate in duct I decreases, but it is increased in duct II. Thus, the stronger the applied magnetic field, the longer the conducting part of the barrier, and the higher the pressure gradient in the duct I, drop the integrated flow rate in the right part of the electrodynamically coupled ducts.

4.1.7 Semi-infinite Strip with Partly Nonconducting and Partly Perfectly Conducting Side Wall

The steady MHD flow of a viscous, incompressible, and electrically conducting fluid in a semi-infinite duct under a horizontally applied magnetic field is considered. The configuration of the domain, displayed in Fig. 4.18, is obtained by extending the top wall of a rectangular duct to infinity so that the Hartmann walls become semi-infinite in length, and kept at the same magnetic field value k but opposite in sign. That is, current enters from one infinite wall and leaves from the other. The wall that connects the two semi-infinite walls is partly nonconducting $B = \pm k$ and partly perfectly conducting for a length 2ℓ in the middle. The semi-infinite walls are discretized by using boundary elements for a finite length due to the regularity condition as $y \to \infty$, and properties of Bessel functions that appear in the fundamental solution for large argument [65], and will be given in Sect. 4.1.8. The basic coupled steady MHD equations obtained from Eq. 4.1 are

$$\left.\begin{array}{l} \nabla^2 V + Ha\dfrac{\partial B}{\partial x} = -1 \\[2ex] \nabla^2 B + Ha\dfrac{\partial V}{\partial x} = 0 \end{array}\right\} \quad \text{in } \Omega\,. \tag{4.5}$$

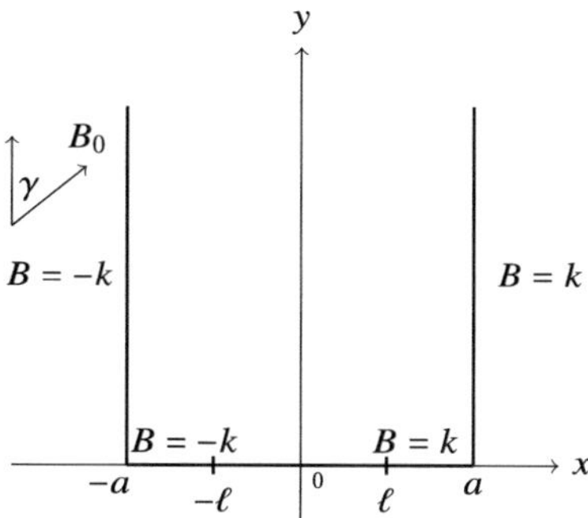

Fig. 4.18 Geometry of the MHD flow in semi-infinite strip

The boundary conditions that are suitable in practice for the MHD flow in an infinite region can be expressed as

$$\begin{aligned}
&V(\pm a, y) = 0 && 0 \le y < \infty \\
&V(x, 0) = 0 && -a < x < a \\
&B(\pm a, y) = \pm k && 0 \le y < \infty \\
&B(x, 0) = -k && -a \le x < -\ell \\
&B(x, 0) = k && \ell < x \le a \\
&\frac{\partial B}{\partial y}(x, 0) = 0 && -\ell \le x \le \ell \\
&|V(x, y)| < \infty, \;\; |B(x, y)| < \infty && \text{as } y \to \infty \\
&|V(x, y)| \to 0, \;\; |B(x, y)| \to 0 \;\; \text{as } y \to \infty \text{ in the vicinity of } x = \pm a
\end{aligned} \tag{4.6}$$

imposing also boundedness conditions

$$\left|\frac{\partial V}{\partial n}\right| < \infty, \; \left|\frac{\partial B}{\partial n}\right| < \infty \quad \text{as } y \to \infty \tag{4.7}$$

for the domain; and on the semi-infinite walls we have

$$\left|\frac{\partial V}{\partial n}\right| \to 0, \; \left|\frac{\partial B}{\partial n}\right| \to 0 \quad \text{as } y \to \infty \text{ in the vicinity of } x = \pm a \tag{4.8}$$

where the conductivity changes abruptly at the points $(-\ell, 0)$ and $(\ell, 0)$.

The BEM results are shown in Fig. 4.19 for increasing values of Ha, and for a constant induced current value $k = 0.5$ on the semi-infinite walls when the side wall at $y = 0$ is partly perfectly conducting for a length of $\ell = 0.4$ symmetrically about the origin. External magnetic field applies horizontally ($\gamma = \pi/2$). It is noticed that as Ha increases, the well-known MHD flow characteristics are established. That is, the flow becomes uniform, and the fluid is stagnant at the center of the duct showing the balance of Lorentz and pressure forces. Boundary layer formations for both the velocity and the induced magnetic field are well observed as Ha increases with

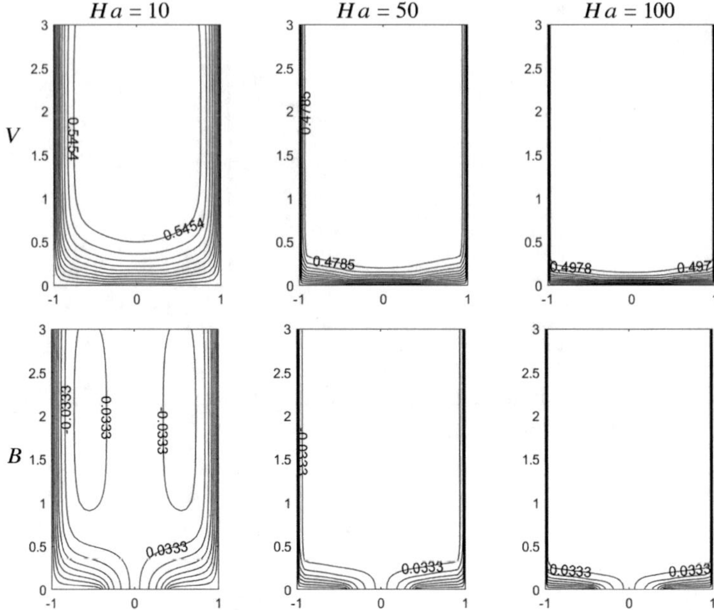

Fig. 4.19 Equi-velocity (top) and current lines (bottom) for $Ha = 10, 50, 100$ (from left to right), $\gamma = \pi/2, \ell = 0.4, k = 0.5$

thicknesses of order $\frac{1}{Ha}$ near the semi-infinite walls and $\frac{1}{\sqrt{Ha}}$ for the velocity near the bottom wall [7, 51, 52]. Induced magnetic field values are in accordance with the sign of thc wall inducecd current value k, which are negative on the left part of the duct and positive on the right, showing the entering and emanating of the current to the duct.

Figure 4.20 displays the effect of increasing wall induced current value k on the equi-velocity and induced magnetic field lines for $Ha = 30$, and perfectly conducting portion of the side wall is of length $\ell = 0.4$. The velocity behavior is not affected much when k is increasing except a slight increase in its magnitude. Induced current lines are scattered through the whole duct for a small value of k, and they form completely boundary layers near the semi-infinite Hartmann walls, entering and leaving the duct symmetrically from the perfectly conducting portion.

4.1.8 Convergence of Integrals on the Infinite Hartmann Walls

In the previous MHD problem defined by Eqs. (4.5)–(4.8), the boundary Γ is an infinite nature due to the semi-infinite Hartmann walls at $x = -a$ and $x = a$ as $y \to \infty$. Thus, the boundary Γ can be written as

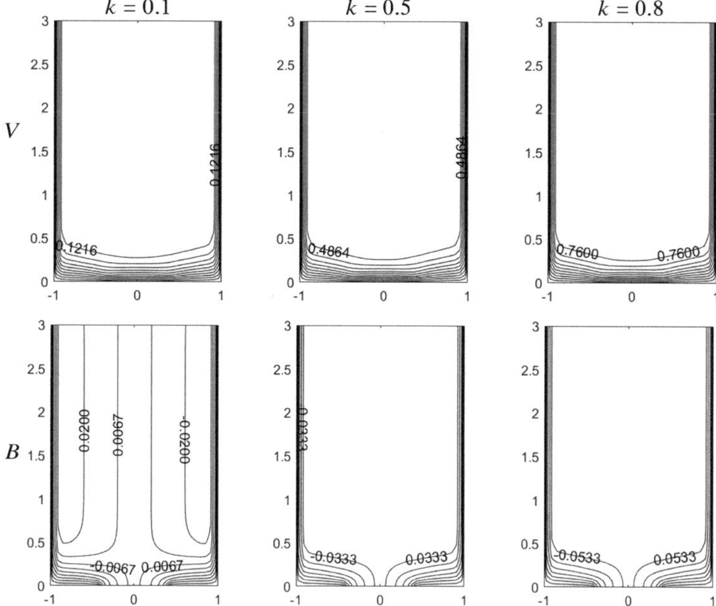

Fig. 4.20 Equi-velocity (top) and current lines (bottom) for $Ha = 30$, $\gamma = \pi/2$, $\ell = 0.4$, $k = 0.1, 0.5, 0.8$ (from left to right)

$$\Gamma = \Gamma_1 + \Gamma\,|_{x=-a} + \Gamma\,|_{x=a}\ ,$$

where $\Gamma_1 = \{(x, 0)\,|-a \le x \le a\}$, and semi-infinite walls contain two parts as $\Gamma\,|_{x=-a} = \Gamma\,\big|_{x=-a,y\ finite} + \Gamma\,\big|_{x=-a,y\to\infty}$, and $\Gamma\,|_{x=a} = \Gamma\,\big|_{x=a,y\ finite} + \Gamma\,\big|_{x=a,y\to\infty}$.

The boundary integrals in the weighted residual form of Eqs. (4.5), which are given in Eqs. (3.10)–(3.11) of Chap. 3, can be restricted only to the boundary $\Gamma = \Gamma_1 + \Gamma\,\big|_{x=-a,y\ finite} + \Gamma\,\big|_{x=a,\ yfinite}$, if the following analysis is adopted [65]:[1]

$$\lim_{y\to\infty}\Bigg[\int_{\Gamma_{x=\pm a}}(V^*\frac{\partial V}{\partial n} - V\frac{\partial V^*}{\partial n})d\Gamma + \int_{\Gamma_{x=\pm a}}(B^*\frac{\partial B}{\partial n} - B\frac{\partial B^*}{\partial n})d\Gamma + \int_{\Gamma_{x=\pm a}} Ha(V^*B + B^*V)n_x d\Gamma\Bigg] = 0 \quad (4.9)$$

or when it is rewritten

[1] Section 4.1.8 was published in International Journal for Numerical Methods in Fluids, Volume 70, C. Bozkaya and M. Tezer-Sezgin, BEM solution to magnetohydrodynamic flow in a semi-infinite duct, 300–312, Copyright Wiley (2012).

$$\lim_{y\to\infty}\left[\int_{\Gamma_{x=\pm a}}(V^*\frac{\partial V}{\partial n}d\Gamma+\int_{\Gamma_{x=\pm a}}B^*\frac{\partial B}{\partial n}d\Gamma+\int_{\Gamma_{x=\pm a}}(HaB^*n_x-\frac{\partial V^*}{\partial n})Vd\Gamma\right.,$$

$$\left.+\int_{\Gamma_{x=\pm a}}(HaV^*n_x-\frac{\partial B^*}{\partial n})Bd\Gamma\right]=0 \tag{4.10}$$

where V^* and B^* replace V_1^*, B_1^* and V_2^*, B_2^* in Eqs. (3.10) and (3.11), respectively.

The replacement of V_1^*, B_1^* or V_2^*, B_2^* from Eq. (3.24) with the integrals containing the terms $\frac{1}{2\pi}K_0(\frac{Ha}{2}r)$ multiplied by either $\cosh(\frac{Ha}{2}r_x)$ or $\sinh(\frac{Ha}{2}r_x)$, and $\frac{1}{2\pi}K_1(\frac{Ha}{2}r)$ multiplied by either $\cosh(\frac{Ha}{2}r_x)\partial r/\partial n$ or $\sinh(\frac{Ha}{2}r_x)\partial r/\partial n$ in $\frac{\partial V_1^*}{\partial n}=\frac{\partial B_2^*}{\partial n}$, $\frac{\partial V_2^*}{\partial n}=\frac{\partial B_1^*}{\partial n}$ (Eq. (3.24)). The first two integrals in (4.10) drop as $y\to\infty$ from the behavior of modified Bessel functions K_0 and K_1 for large arguments and the behavior of exponential function.

Considering the first integral in (4.10)

$$\begin{aligned}\lim_{y\to\infty}\int_{\Gamma_{x=\pm a}}V^*\frac{\partial V}{\partial n}d\Gamma &= \lim_{y\to\infty}\int_{\Gamma_{x=\pm a}}\frac{1}{2\pi}K_0(\frac{Ha}{2}r)\cosh(\frac{Ha}{2}r_x)\frac{\partial V}{\partial n}d\Gamma\\ &= \lim_{y\to\infty}\frac{1}{2\pi}\int_{\Gamma_{x=\pm a}}\frac{1+e^{-Har_x}}{2}e^{\frac{Ha}{2}r_x}\sqrt{\frac{\pi}{2}}\frac{e^{-\frac{Ha}{2}r}}{\sqrt{r}\sqrt{\frac{Ha}{2}}}\frac{\partial V}{\partial n}d\Gamma\\ &= \lim_{y\to\infty}\frac{1}{4\sqrt{Ha\pi}}\int_{\Gamma_{x=\pm a}}(1+e^{-Har_x})\frac{e^{-\frac{Ha}{2}(r-r_x)}}{\sqrt{r}}\frac{\partial V}{\partial n}d\Gamma\,.\end{aligned} \tag{4.11}$$

Because r is the distance from the source point to the field point (both of these points are any two nodes on the boundary) and r_x is the x-component of the vector $\mathbf{r}$, we encounter three cases when $y\to\infty$ at $x=\pm a$. These are:

1. $r\to\infty$.
2. r is large but bounded.
3. r is small (finite).

In the first case, the term $e^{-\frac{Ha}{2}(r-r_x)}\to 0$ and the integral (4.11) drops. In the second case, both $r-r_x$ and $e^{-\frac{Ha}{2}(r-r_x)}$ are bounded. However, $1/\sqrt{r}\to 0$ uniformly, which leads to the drop of the integral (4.11). On the other hand, although $e^{-\frac{Ha}{2}(r-r_x)}/\sqrt{r}$ has a positive value when r is small (finite), the integral over $\Gamma_x=\pm a$ still drops because $\partial V/\partial n\to 0$ as $y\to\infty$ in the vicinity of $x=\pm a$.

Similarly, the second integral in Eq. (4.10) tends to zero as $y\to\infty$ at $x=\pm a$ since the behavior of V^*, $\partial V/\partial n$ and B^*, $\partial B/\partial n$ are the same, respectively.

Now, consider the third integral in Eq. (4.10). This integral can be written as

$$
\begin{aligned}
&\lim_{y\to\infty}\int_{\Gamma_{x=\pm a}} (HaB^*n_x - \frac{\partial V^*}{\partial n})Vd\Gamma \\
&= \lim_{y\to\infty}\int_{\Gamma_{\pm a}} \left[\frac{Ha}{2\pi}K_0(\frac{Ha}{2}r)\sinh(\frac{Ha}{2}r_x)n_x\right. \\
&-\frac{Ha}{4\pi}K_1(\frac{Ha}{2}r)\cosh(\frac{Ha}{2}r_x)\left(\frac{r_x}{r}n_x+\frac{r_y}{r}n_y\right) \\
&\left.-\frac{Ha}{4\pi}K_0(\frac{Ha}{2}r)\sinh(\frac{Ha}{2}r_x)n_x\right]Vd\Gamma \\
&= \lim_{y\to\infty}\frac{\sqrt{Ha}}{8\sqrt{\pi}}\int_{\Gamma_{x=\pm a}}\left[(1+e^{-Har_x})\frac{e^{-\frac{Ha}{2}(r-r_x)}}{\sqrt{r}}\frac{r_x}{r}\right. \\
&\left.-(1-e^{-Har_x})\frac{e^{-\frac{Ha}{2}(r-r_x)}}{\sqrt{r}}\right]Vd\Gamma
\end{aligned}
\tag{4.12}
$$

since $n_x = \pm 1$ and $n_y = 0$ at $x = \pm a$.

Similar arguments as in the first and second integrals in Eq. (4.10) are also held here. There exist also three cases. In the first case, $r \to \infty$ and thus $r - r_x \to \infty$, and $e^{-\frac{Ha}{2}(r-r_x)} \to 0$ and the integral (4.12) vanishes. In the second case, r is large but bounded and $r - r_x$ is also bounded. However, $1/\sqrt{r} \to 0$ uniformly, dropping integral (4.12). On the other hand, in the third case, r is small (finite) and in the same order with r_x, that is $\frac{r_x}{r} \approx O(1)$. Thus, the integral in (4.12) takes the form

$$\lim_{y\to\infty}\frac{\sqrt{Ha}}{4\sqrt{\pi}}\int_{\Gamma_{x=\pm a}}\frac{e^{-\frac{Ha}{2}(r+r_x)}}{\sqrt{r}}Vd\Gamma$$

and $e^{-\frac{Ha}{2}(r+r_x)}/\sqrt{r}$ has a positive value. However, this integral vanishes since $|V| \to 0$ in the vicinity of $x = \pm a$ as $y \to \infty$.

Finally, the last integral in (4.10) also drops with the similar behavior of V^* and B^*, $\partial V^*/\partial n$ and $\partial B^*/\partial n$, V and B, at $x = \pm a$, $y \to \infty$.

Thus, the boundary integrals in Eqs. (3.10) and (3.11) are reduced only to the boundary $\Gamma = \Gamma_1 + \Gamma\big|_{x=-a,y\ finite} + \Gamma\big|_{x=a,y\ finite}$.

4.2 Electrically Driven MHD Flow in Infinite Channels

In Sect. 4.1 the MHD duct flows that are pressure driven are considered. The present section is devoted to MHD channel flows that are driven by imposed electric currents. The flow regions are the infinite regions, either the upper half plane or the region between two parallel infinite plates. The applied magnetic field is perpendicular to the x-axis ($\gamma = 0$).

4.2.1 MHD Flow in the Upper Half Plane

MHD flow equations in the upper half plane have been solved in the work of Sezgin [66] by reducing the MHD flow equations to dual integral equations, and then to the solution of a Fredholm integral equation of the second kind. In this chapter, the kernel contains infinite integrals that give difficulties to compute. Later, Tezer-Sezgin and Bozkaya [67] presented BEM solutions solving the equations in coupled form as a whole with the fundamental solution they have derived for coupled MHD equations.

The partial differential equations describing electrically driven flows are the same as those MHD duct flows where pressure gradient $\frac{\partial p}{\partial z}$ is taken as zero [7, 13] and they are given in nondimensional form as

$$\left.\begin{array}{l} \nabla^2 V + Ha\dfrac{\partial B}{\partial y} = 0 \\[2ex] \nabla^2 B + Ha\dfrac{\partial V}{\partial y} = 0 \end{array}\right\} \quad \text{in } \Omega\,, \tag{4.13}$$

where Ω is the upper half plane.

The general form of the boundary conditions that are suitable in practice for the MHD flow in an infinite region can be expressed as

$$\begin{array}{ll} V(x,0) = 0 & -\infty < x < \infty \\ B(x,0) = \bar{B} & \text{on } \Gamma_I \\ \dfrac{\partial B}{\partial y}(x,0) = 0 & \text{on } \Gamma_C \\ |V(x,y)| < \infty, \ |B(x,y)| < \infty & \text{as } x^2 + y^2 \to \infty \\ V \to 0, \ B \to 0 & \text{as } x \to -\infty \\ \left|\dfrac{\partial V}{\partial n}\right| < \infty, \ \left|\dfrac{\partial B}{\partial n}\right| < \infty & \text{as } x^2 + y^2 \to \infty \\[2ex] \left|\dfrac{\partial V}{\partial n}\right| \to 0, \ \left|\dfrac{\partial B}{\partial n}\right| \to 0 & \text{as } y \to \infty \text{ in the vicinity of discontinuity points,} \end{array} \tag{4.14}$$

where $\Gamma = \Gamma_I + \Gamma_C$ is the whole x-axis with $\Gamma_I \cap \Gamma_C = \phi$. Γ_I and Γ_C are the nonconducting and perfectly conducting parts of the boundary Γ, respectively. The points where the conductivity changes abruptly are called the discontinuity points and $\bar{B}$ is a known constant value for the induced magnetic field.

4.2.2 Convergence of Infinite Boundary Integrals

BEM formulation of Eqs. (4.13) is given in Eqs. (3.10)–(3.11) (Chap. 3) when the right-hand sides are taken as zero due to the missing pressure gradient term in Eq. (4.13), and external magnetic field is applied vertically.

Since the region in the MHD flow problem (4.2.1) is the upper half plane, the boundary in the integrals in Eqs. (3.10)–(3.11) contains the fictitious boundary Γ_∞ that is the infinitely distant upper semicircle with radius R and the center at the origin. Thus, Eqs. (3.10)–(3.11) can be written containing boundary integrals on both Γ_x and fictitious boundary Γ_∞ where Γ_x is the diameter of the semicircle on the x-axis [67].[2]

So, the boundary integrals in (3.10)–(3.11) can be restricted only to the boundary Γ_x, if the following condition is obeyed:

$$\lim_{R\to\infty}\left[\int_{\Gamma_\infty}(V^*\frac{\partial V}{\partial n}-V\frac{\partial V^*}{\partial n})d\Gamma_\infty+\int_{\Gamma_\infty}(B^*\frac{\partial B}{\partial n}-B\frac{\partial B^*}{\partial n})d\Gamma_\infty\right.$$
$$\left.+\int_{\Gamma_\infty}Ha(V^*B+B^*V)n_y d\Gamma_\infty\right]=0 \tag{4.15}$$

or when it is rewritten

$$\lim_{R\to\infty}\left[\int_{\Gamma_\infty}(V^*\frac{\partial V}{\partial n}d\Gamma_\infty+\int_{\Gamma_\infty}B^*\frac{\partial B}{\partial n}d\Gamma_\infty+\int_{\Gamma_\infty}(HaB^*n_y-\frac{\partial V^*}{\partial n})Vd\Gamma_\infty\right.$$
$$\left.+\int_{\Gamma_\infty}(HaV^*n_y-\frac{\partial B^*}{\partial n})Bd\Gamma_\infty\right]=0, \tag{4.16}$$

where again V^* and B^* replace V_1^*, B_1^* and V_2^*, B_2^* in Eqs. (3.10) and (3.11), respectively.

The replacement of V_1^*, B_1^* or V_2^*, B_2^* and also their normal derivatives that are given in Eq. (3.24) results in the integrals containing the terms $\frac{1}{2\pi}K_0(\frac{Ha}{2}r)$ multiplied by either $\cosh(\frac{Ha}{2}r_y)$ or $\sinh(\frac{Ha}{2}r_y)$, and $\frac{1}{2\pi}K_1(\frac{Ha}{2}r)$ multiplied by either $\cosh(\frac{Ha}{2}r_y)\frac{\partial r}{\partial n}$ or $\sinh(\frac{Ha}{2}r_y)\frac{\partial r}{\partial n}$. The first two integrals in (4.16) drop as $R\to\infty$ $(r\to\infty)$ from the behavior of modified Bessel functions K_0 and K_1 for large arguments and the behavior of exponential function, as explained below.

Considering the first integral in (4.16)

[2] Section 4.2.2 was published in Journal of Computational and Applied Mathematics, Volume 225(2), M. Tezer-Sezgin and C. Bozkaya, The boundary element solution of the magnetohydrodynamic flow in an infinite region, 510–521, Copyright Elsevier (2009).

$$\lim_{R\to\infty}\int_{\Gamma_\infty} V_1^* \frac{\partial V}{\partial n} d\Gamma_\infty = \lim_{R\to\infty}\int_{\Gamma_\infty} \frac{1}{2\pi} K_0(\frac{Ha}{2}r)\cosh(\frac{Ha}{2}r_y)\frac{\partial V}{\partial n}\, d\Gamma_\infty$$

$$= \lim_{R\to\infty}\frac{1}{2\pi}\int_{\Gamma_\infty} \frac{1+e^{-Har_y}}{2} e^{\frac{Ha}{2}r_y}\sqrt{\frac{\pi}{2}}\frac{e^{-\frac{Ha}{2}r}}{\sqrt{r}\sqrt{\frac{Ha}{2}}}\frac{\partial V}{\partial n}\, d\Gamma_\infty$$

$$= \lim_{R\to\infty}\frac{1}{4\sqrt{Ha\pi}}\int_{\Gamma_\infty} (1+e^{-Har_y})\frac{e^{-\frac{Ha}{2}(r-r_y)}}{\sqrt{r}}\frac{\partial V}{\partial n}\, d\Gamma_\infty\,. \tag{4.17}$$

If $(r - r_y) \to \infty$, the integral drops obviously. If $(r - r_y)$ is bounded, so is $e^{-\frac{Ha}{2}(r-r_y)}$ bounded (except near y-axis) but $\frac{1}{\sqrt{r}} \to 0$ uniformly, which leads to the drop of the integral. However, when x approaches zero in a narrow segment around the y-axis, the integral of the terms $(e^{-\frac{Ha}{2}(r-r_y)})/\sqrt{r}$ has a positive value. But the integral over Γ_∞ still drops since $\frac{\partial V}{\partial n} \to 0$ as $x \to 0,\ y \to \infty$. Similarly, the second integral in (4.16) tends to zero as $R \to \infty$.

Also, the third integral in (4.16) can be written as

$$\lim_{R\to\infty}\int_{\Gamma_\infty}(HaB_1^* n_y - \frac{\partial V_1^*}{\partial n})V d\Gamma_\infty$$

$$= \lim_{R\to\infty}\int_{\Gamma_\infty}\left[\frac{Ha}{2\pi}K_0(\frac{Ha}{2}r)\sinh(\frac{Ha}{2}r_y)n_y \right.$$
$$-\frac{Ha}{4\pi}K_1(\frac{Ha}{2}r)\cosh(\frac{Ha}{2}r_y)\left(\frac{r_x}{r}n_x+\frac{r_y}{r}n_y\right)$$
$$\left.-\frac{Ha}{4\pi}K_0(\frac{Ha}{2}r)\sinh(\frac{Ha}{2}r_y)n_y\right]V d\Gamma_\infty \quad . \tag{4.18}$$

$$= \lim_{R\to\infty}\frac{\sqrt{Ha}}{8\sqrt{\pi}}\int_{\Gamma_\infty}\left[(1+e^{-Har_y})\frac{e^{-\frac{Ha}{2}(r-r_y)}}{\sqrt{r}}\frac{r_y}{r}\right.$$
$$\left.-(1-e^{-Har_y})\frac{e^{-\frac{Ha}{2}(r-r_y)}}{\sqrt{r}}\right]V d\Gamma_\infty$$

Similar arguments as in the case of first and second integrals hold for $r - r_y \to \infty$ and $r - r_y$ is bounded. However again near the y-axis, r_y and r are of the same order, i.e., $\frac{r_y}{r} \approx O(1)$. Thus, the integral in (4.18) takes the form $\lim_{R\to\infty}\frac{\sqrt{Ha}}{4\sqrt{\pi}}\int_{\Gamma_\infty}\frac{e^{-\frac{Ha}{2}(r+r_y)}}{\sqrt{r}}V d\Gamma_\infty$, which tends to zero as $R \to \infty$ since V is bounded. Similarly, the last integral in (4.16) also drops with the behaviors of V^* and $\partial B^*/\partial n$ as $R \to \infty$.

Thus, the boundary integrals in (3.10) or (3.11) can be restricted only to the boundary Γ_x, which is the diameter of the semicircle on the x-axis.

4.2.3 MHD Flow in the Upper Half Plane with Perfectly Conducting Left and Nonconducting Right Walls

We assume that the flow is laminar, steady and the fluid is incompressible. An external circuit is connected so that a current enters the fluid at (0, 0) and leaves it at infinity. The flow is driven by the interaction of vertically imposed electric currents and the uniform, transverse magnetic field. Thus, this is a simple MHD flow problem on the upper half plane ($y \geq 0$) defined by Eq. (4.13) with the boundary conditions [67]

$$\begin{aligned}
&V(x,0) = 0 && -\infty < x < \infty \\
&B(x,0) = 1 && 0 \leq x < \infty \\
&\frac{\partial B}{\partial y}(x,0) = 0 && -\infty < x < 0 \\
&V \to 0,\ B \to 0 && \text{as } x \to -\infty \\
&\left|\frac{\partial V}{\partial n}\right| \to 0,\ \left|\frac{\partial B}{\partial n}\right| \to 0 && \text{as } x \to 0,\ y \to \infty \\
&|V| < \infty,\ |B| < \infty && \text{as } x^2 + y^2 \to \infty \\
&\left|\frac{\partial V}{\partial n}\right| < \infty,\ \left|\frac{\partial B}{\partial n}\right| < \infty && \text{as } x^2 + y^2 \to \infty\,.
\end{aligned} \tag{4.19}$$

The wall $y = 0$ is partly nonconducting but kept at a constant current, $B = 1$ ($x > 0$), and partly perfectly conducting, $\partial B/\partial y = 0$ ($x < 0$), as shown in Fig. 4.21.

In Fig. 4.22, the equal velocity and current lines are plotted for this simple MHD problem when $Ha = 10, 30, 100$ in the region $-2 \leq x \leq 2$ and $0 \leq y \leq 4$ that

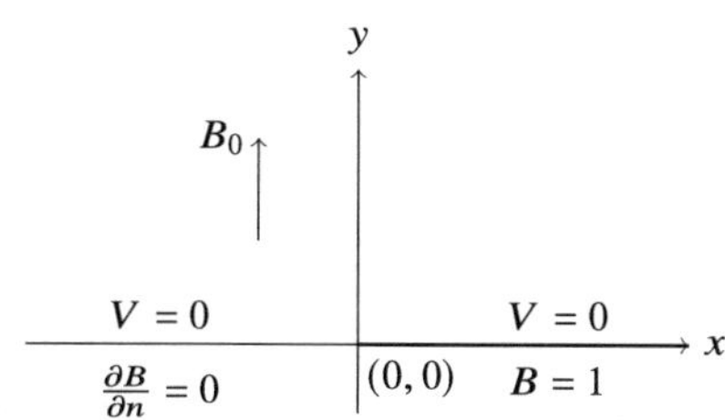

Fig. 4.21 Geometry of the MHD flow in the upper half plane

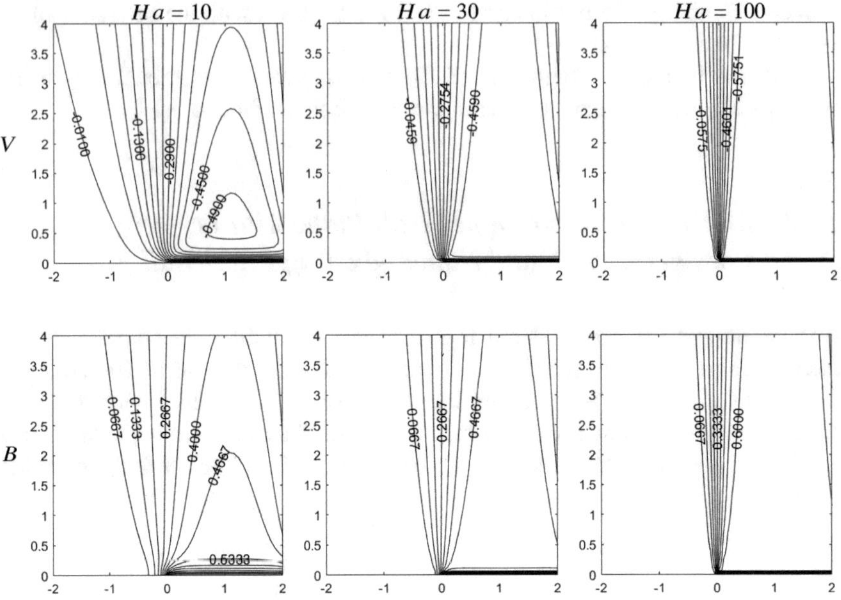

Fig. 4.22 Equi-velocity (top) and current lines (bottom) for $Ha = 10, 30, 100$ (from left to right), $\gamma = 0$

is believed to show the behaviors of the flow and induced currents in the infinite upper plane. As Ha increases, boundary layer formation starts taking place near the nonconducting boundary ($y = 0$, $x \geq 0$) for both the velocity and the induced magnetic field. An increase in Ha causes stagnant regions for the velocity field in the regions except the parabolic boundary layer emanating from the point of discontinuity (i.e., the origin) and propagating in the upper half plane. Induced magnetic field lines obey the $\partial B/\partial n = 0$ condition on the left boundary ($x < 0$), but they concentrate to the origin as Ha increases. The thickness of parabolic boundary layer emanating from the origin has been computed by Bozkaya and Tezer-Sezgin [67] and given in the following section.

4.2.4 The Thickness of the Parabolic Boundary Layer

It is known that there is a boundary layer (Hartmann layer) near the insulated (or nonconducting) walls that is perpendicular to the applied magnetic field, Hunt [52], of order $1/Ha$. Thus, on the portion of the x-axis where the wall is nonconducting ($x \geq 0$), we have Hartmann layers of order $1/Ha$, and on the conducting portion, the flow is almost stagnant, and there are hardly any current lines. This behavior is depicted in Fig. 4.23, which gives the boundary layer thickness versus the Hartmann number Ha compared with the function $1/Ha$.

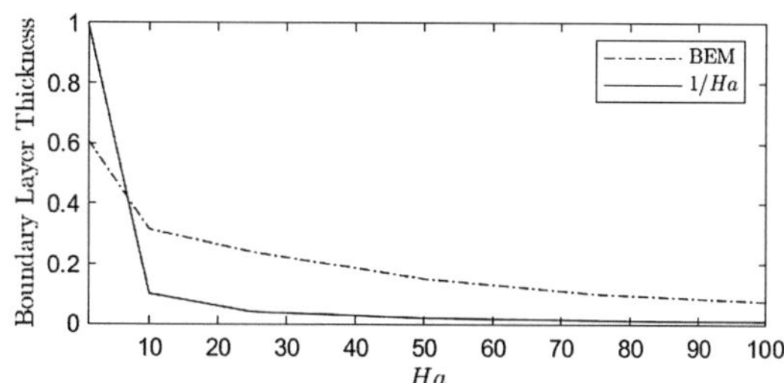

Fig. 4.23 Variation of Hartmann boundary layer thickness

The system of BEM discretized equations for Eq. (4.13) is given in the system (3.28) of Chap. 3, and entries of the matrices $\mathbf{H}$, $\mathbf{G}$, $\mathbf{H}^1$, $\mathbf{G}^1$ are given in Eq. (3.31).

The integrals in the entries h_{ij}, g_{ij} and h^1_{ij}, g^1_{ij} contain the product of Bessel and hyperbolic functions as

$$\int_{\Gamma_j} K_0(\frac{Ha}{2}r)\sinh(\frac{Ha}{2}r_y)d\Gamma_j \quad , \quad \int_{\Gamma_j} K_0(\frac{Ha}{2}r)\cosh(\frac{Ha}{2}r_y)d\Gamma_j$$

and

$$\int_{\Gamma_j} K_1(\frac{Ha}{2}r)\sinh(\frac{Ha}{2}r_y)\frac{\partial r}{\partial n}d\Gamma_j \, , \quad \int_{\Gamma_j} K_1(\frac{Ha}{2}r)\cosh(\frac{Ha}{2}r_y)\frac{\partial r}{\partial n}d\Gamma_j \, , \tag{4.20}$$

where $\mathbf{r} = (r_x, r_y)$ is the distance vector between the source and field points.

Since the integrations are on the boundary elements Γ_j along the x-axis, we have $d\Gamma_j = dx$.

Each integral in (4.20) can be approximated for $r_y >> r_x$ as [67][3]

$$\begin{aligned} &\frac{1 \pm e^{-Har_y}}{2}\frac{\sqrt{\pi}}{\sqrt{Ha}}\int_{\Gamma_j}\frac{e^{\frac{-Ha}{2}r_y(\sqrt{1+r_x^2/r_y^2}-1)}}{(r_x^2+r_y^2)^{\frac{1}{4}}}dx \\ &= (1 \pm e^{-Har_y})\frac{\sqrt{\pi}}{2\sqrt{Ha}}\int_0^{|x|}\frac{e^{\frac{-Har_x^2}{4r_y}}}{\sqrt{r_y}}dx \end{aligned} \tag{4.21}$$

since we consider the boundary layer close to discontinuity points, i.e., in the regions $|x|$ for small r_x. With the change of variable $u = \sqrt{Ha}r_x/(2\sqrt{r_y})$, it can still be transformed to

[3] Section 4.2.4 was published in Journal of Computational and Applied Mathematics, Volume 225(2), M. Tezer-Sezgin and C. Bozkaya, The boundary element solution of the magnetohydrodynamic flow in an infinite region, 510–521, Copyright Elsevier (2009).

$$\frac{\pi}{2Ha}(1 \pm e^{-Har_y})\, erf(\frac{\sqrt{Ha}\, r_x}{2\sqrt{r_y}}), \tag{4.22}$$

and with the help of the property of error function $erf(x)$ for $0 \leq x < \infty$ (rational approximation), it can be written in the form

$$\frac{\pi}{2Ha}(1 \pm e^{-Har_y})\left[1 - \frac{a_1}{1+px} e^{-\frac{Har_x^2}{4r_y}} + \cdots\right],$$

$$= \frac{\pi}{2Ha}(1 \pm e^{-Har_y})\left[1 - a_1\left(1 - p\frac{\sqrt{Ha}|r_x|}{2\sqrt{r_y}} \cdots\right) e^{-\frac{Har_x^2}{4r_y}} + \cdots\right] \tag{4.23}$$

where a_1 and p are positive constants.

Since r_x and r_y behave like x and y, respectively, the thickness of the boundary layer that emanates from the points of discontinuities of boundary conditions on the x-axis and lies on the upper half of the plane $y > 0$ is obtained as

$$|x| = 2\frac{\sqrt{y}}{\sqrt{Ha}}. \tag{4.24}$$

This result is also in accordance with the thickness of the secondary layer on the boundary parallel to the applied magnetic field mentioned by Hunt [52] that is of order $1/\sqrt{Ha}$. It shows that this type of secondary layer (parabolic layer) also appears from the points of discontinuity in boundary conditions. Figure 4.24 also shows the variation of the thickness of this parabolic boundary layer versus Ha compared with the function $1/\sqrt{Ha}$.

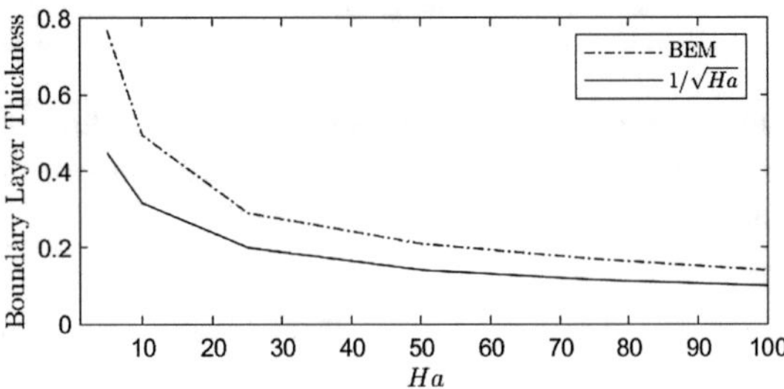

Fig. 4.24 Variation of parabolic boundary layer thickness

4.2.5 MHD Flow in the Upper Half Plane with Partly Nonconducting and Partly Perfectly Conducting Wall

The MHD flow on the upper half of an infinite plate for the case when the flow is driven by the current produced by an electrode of length 2ℓ, placed in the middle of the plate, therefore conducting, is considered. Imposed electric currents enter the fluid at $x = \pm\ell$ through external circuits and move up on the plane. The partial differential equations describing the flow are the same as in Sect. 4.2.3 (Eq. (4.13)) with the boundary conditions (see Fig. 4.25) [67]

$$\begin{aligned}
&V(x,0) = 0 && -\infty < x < \infty \\
&B(x,0) = -1 && -\infty < x < -\ell \\
&B(x,0) = 1 && \ell < x < \infty \\
&\frac{\partial B}{\partial y}(x,0) = 0 && -\ell < x < \ell \\
&\left|\frac{\partial V}{\partial n}\right| \to 0,\ \left|\frac{\partial B}{\partial n}\right| \to 0 && \text{as } x \to \pm\ell,\ y \to \infty \\
&|V| < \infty,\ |B| < \infty && \text{as } x^2 + y^2 \to \infty \\
&\left|\frac{\partial V}{\partial n}\right| < \infty,\ \left|\frac{\partial B}{\partial n}\right| < \infty && \text{as } x^2 + y^2 \to \infty\,.
\end{aligned} \tag{4.25}$$

Infinite integrals are computed at the finite intervals $x \in [-2, 2]$ and $y \in [0, 4]$ due to the analysis done for the restriction of infinite integrals to the diameter of the semicircle of radius 2. Figures 4.26 and 4.27 display the equal velocity and magnetic field lines as Ha and the length of the conducting portion increase, respectively. As Ha increases, boundary layers are formed near $y = 0$ line for $x > \ell$ and $x < -\ell$ for both the velocity and the induced magnetic field. Except the narrow regions near $y = 0$ and $x = 0$, the velocity is almost constant and the fluid is stagnant. For $x < \ell$ and $x > -\ell$, and for small values of y, the current lines are perpendicular to the x-axis owing to the boundary condition $\frac{\partial B}{\partial n} = 0$. Further, in most parts of the region, the value of the induced magnetic field is stationary. An increase in the length of conducting portion ℓ also develops a stagnant region in front of the conducting wall.

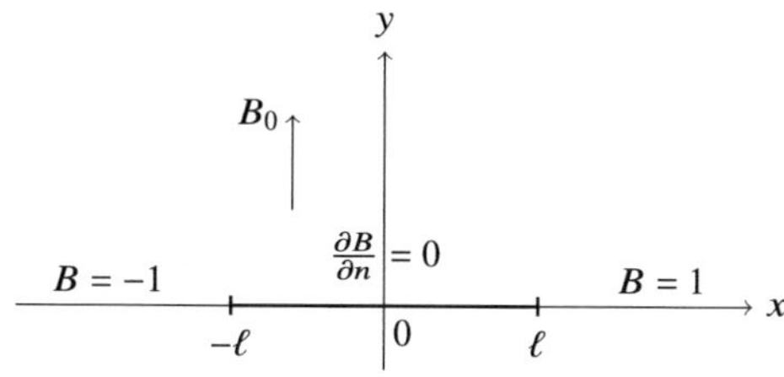

Fig. 4.25 Geometry of the MHD flow in the upper half plane with partly perfectly conducting–partly nonconducting wall

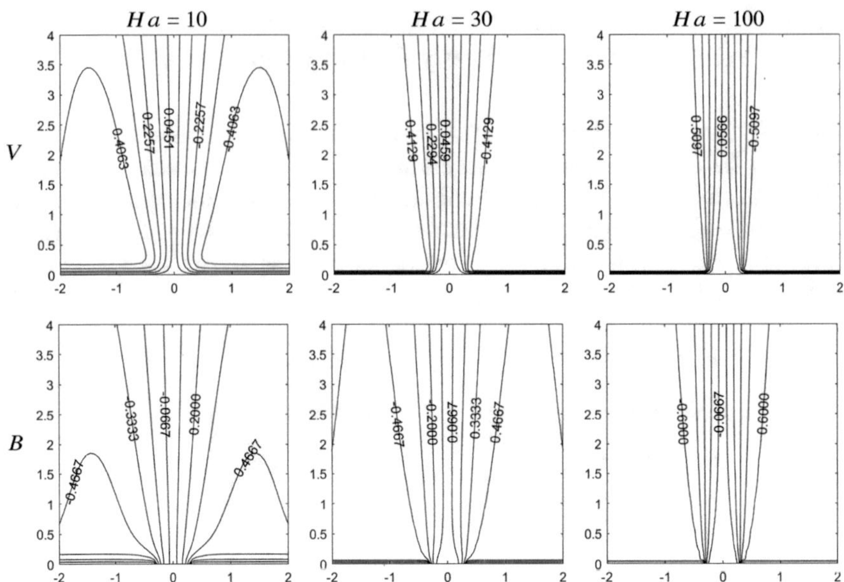

Fig. 4.26 Equi-velocity (top) and current lines (bottom) for $Ha = 10, 30, 100$ (from left to right), $\ell = 0.3, \gamma = 0, k = \pm 1$

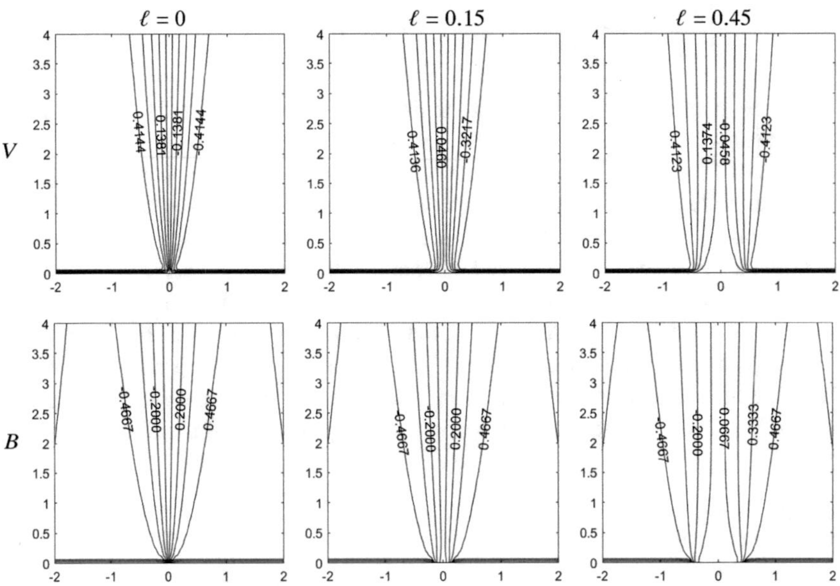

Fig. 4.27 Equi-velocity (top) and current lines (bottom) for $\ell = 0, 0.15, 0.45$ (from left to right), $Ha = 30, \gamma = 0, k = \pm 1$

We notice once again the development of the parabolic boundary layer near $x = 0$. For small values of ℓ and/or Ha, this boundary layer interacts with the similar layer originating at $x = -\ell$. For higher values of ℓ and Ha, however, the two layers are separated [67].

4.2.6 MHD Flow Driven by Electrodes Between Infinite Parallel Plates

The MHD flow between two parallel nonconducting infinite planes extending along the x-axis is investigated for the case when the flow is driven by the current produced by electrodes of length 2ℓ, placed one in each plane, the applied magnetic field being perpendicular to the planes. The rest of the plates are kept at constant current values $B = \pm k$ asymmetrically around the origin. The electrodes are connected to an external circuit so that an electric current travels between the electrodes entering the fluid from one plane and leaving from the other. This problem is the special case of rectangular duct problem where the top and bottom walls are extended to infinity. The configuration of the flow region is shown in Fig. 4.28.

The equations describing such flows are the same as those of MHD duct flow with zero pressure gradient given in Eq. (4.13). The velocity and the induced magnetic field are in the z-direction (axis of the channel) and of functions of the plane coordinates x and y again. The supplied boundary conditions are [68]

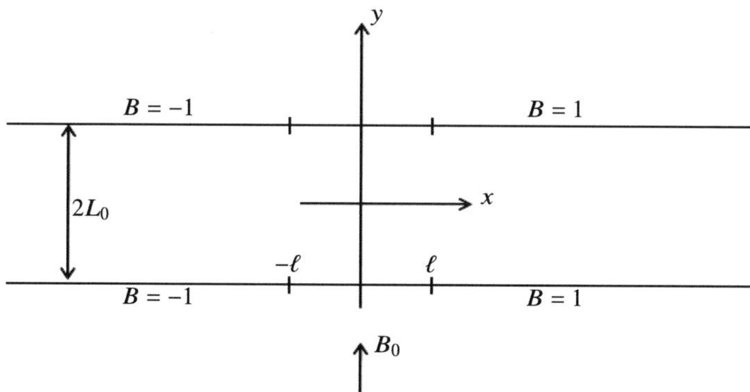

Fig. 4.28 Geometry of MHD flow between two parallel planes

$$
\begin{aligned}
&V(x,-L_0)=V(x,L_0)=0 && -\infty\le x<\infty\\
&B(x,\pm L_0)=1 && l<x<\infty\\
&B(x,\pm L_0)=-1 && -\infty<x<-l\\
&\frac{\partial B}{\partial y}(x,\pm L_0)=0 && -l\le x\le l\\
&|V(x,y)|<\infty,\ \ |B(x,y)|<\infty && \text{as } x\to\pm\infty\\
&|V(x,y)|\to 0,\ \ |B(x,y)|\to 0 && \text{as } x\to\pm\infty \text{ in the vicinity of } y=\pm L_0\\
&\left|\frac{\partial V}{\partial n}\right|<\infty,\ \ \left|\frac{\partial B}{\partial n}\right|<\infty && \text{as } x\to\pm\infty\\
&\left|\frac{\partial V}{\partial n}\right|\to 0,\ \ \left|\frac{\partial B}{\partial n}\right|\to 0 && \text{as } x\to\pm\infty \text{ in the vicinity of } y=\pm L_0.
\end{aligned}
\tag{4.26}
$$

The BEM formulation of Eq. (4.13) is given in Eqs. (3.10)–(3.11) of Chap. 3 with zero right-hand sides since there is no pressure gradient.

The boundary of the problem contains finite and infinite parts as $\Gamma = \Gamma_{y=\pm L_0,\, x\ finite} + \Gamma_{y=\pm L_0,\, x\to\infty}\}$. The boundary integrals in Eqs. (3.10)–(3.11) where n_x is replaced by n_y can be restricted to $\Gamma = \Gamma_{y=\pm L_0,\, x\ finite}$ since the infinite part integrals drop with a similar argument carried in Sect. 4.1.8 for semi-infinite Hartmann walls and Sect. 4.2.2 for the upper half plane. Infinite integrals are also obtained as $x \to \pm\infty$ where $n_x = 0$ at $y = \pm L_0$ but $n_y = \pm 1$. The term $e^{\frac{Ha}{2}(r-r_y)} \to 0$ as $r \to \infty$, and $\frac{1}{\sqrt{r}} \to 0$ uniformly when r is large but bounded since $e^{\frac{Ha}{2}(r-r_y)}$ is bounded. For small values of r both $|\frac{\partial V}{\partial n}| \to 0$, $|\frac{\partial B}{\partial n}| \to 0$ and $|V| \to 0$, $|B| \to 0$ hold in the vicinity of $y = \pm L_0$ as $x \to \pm\infty$.

Thus, all the infinite integrals drop, and Γ is restricted to finite part of the boundary in Eqs. (3.10)–(3.11) containing n_y instead of n_x.

In the computations, this finite part is taken as the interval $[-1.75, 1.75]$ on the x-axis. We present equal velocity and current lines for increasing length of the electrodes for $Ha = 10$ and $Ha = 50$ in Fig. 4.29 for $L_0 = 1$. One can notice that, although the problem is solved for an infinite strip, the flow is confined to a relatively small region near the lines $x = \pm\ell$. In the rest of the region, the fluid is almost stagnant. Boundary layer formation for the velocity starts for increasing values of Ha and for $x \ge \ell$ and $x \le -\ell$ as is observed in the other infinite plane problems containing conducting parts. Induced current lines are perpendicular to $y = \pm 1$ lines again for $-\ell < x < \ell$ demonstrating the derivative condition on B ($\frac{\partial B}{\partial y} = 0$). Increase in the length of the electrodes causes the formation of a stagnant region for the velocity in front of the conducting part. When $\ell = 0$, this problem reduces to the special case of the flow induced by line electrodes at $x = 0$ set in insulating planes at $y = \pm 1$, with a magnetic field applied in the y-direction. This special case has been considered by Hunt and Williams [69] using asymptotic solution for large Ha in separate regions. The BEM solution has been given together

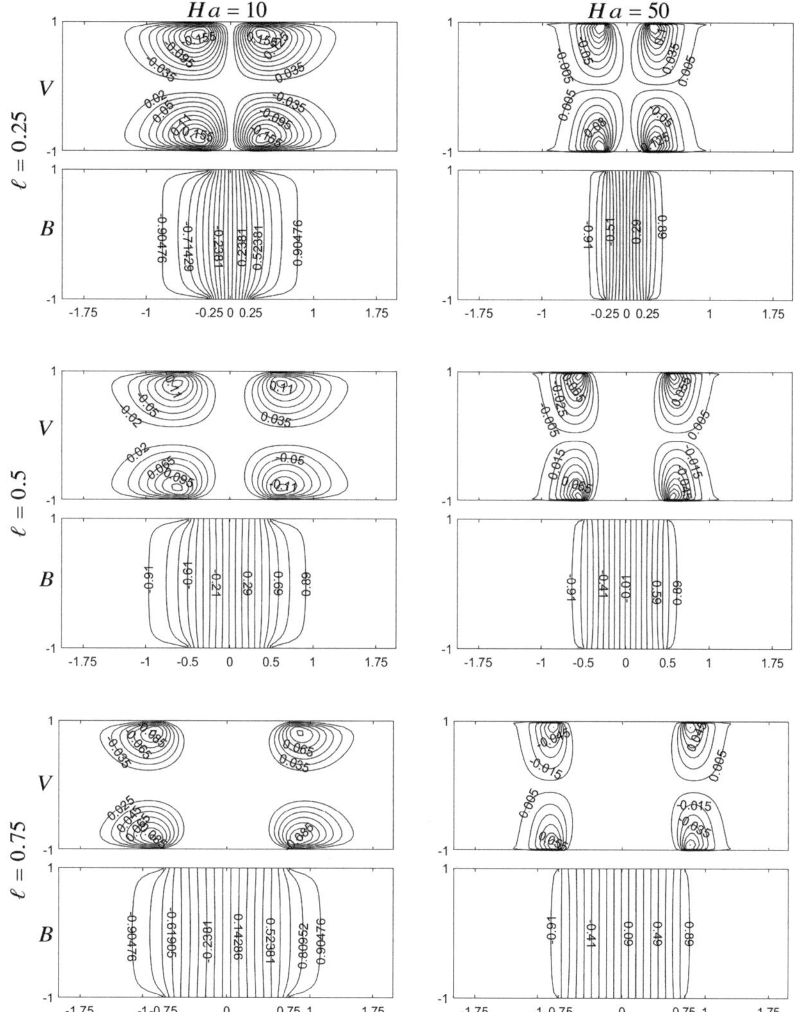

Fig. 4.29 Equi-velocity and current lines for $Ha = 10$ (left) and $Ha = 50$ (right), $\gamma = 0$, $\ell = 0.25, 0.5, 0.75$

with convergence of infinite integrals and thickness of boundary layer by Bozkaya and Tezer-Sezgin in [68].

This chapter closes with the BEM solutions of pressure driven or electrically driven MHD flows in rectangular or circular ducts and in infinite regions. The coupled MHD equations in terms of the velocity and the induced magnetic field are treated as a whole with the fundamental solution of coupled MHD equations. For the combinations of conducting and insulating walls as well as partly insulated–

partly perfectly conducting one wall cases, the numerical results are obtained for several values of problem parameters.

In the next chapter, the DRBEM is going to be used with the fundamental solution of Laplace equation for solving the MHD convection, full MHD, unsteady MHD, and inductionless MHD flow equations iteratively.

Chapter 5
DRBEM Solutions of Unsteady MHD Convection and Inductionless Flows

This chapter contains the dual reciprocity boundary element solutions of MHD flow problems in which the time derivatives (unsteady MHD flow) or the nonlinearity due to the convection terms in the flow and induced current equations are involved. These are the cases where the fundamental solution to the whole differential equation is not available. Thus, all the terms other than Laplacian are treated as nonlinearity, and the fundamental solution of Laplace equation is employed, which is the DRBEM technique. Natural convection MHD flow in cavities is also solved by using the DRBEM.

The interaction between the conducting fluid and an applied magnetic field causes an electromotive force resulting in induced currents that generate an induced magnetic field according to Ampere's Law. In most of the studies in this chapter, the induced magnetic field is neglected due to the small magnetic Reynolds number. In consideration of the induced magnetic field, the governing equations that consist of combination of Navier–Stokes and Maxwell's equations will also contain the induction equations, and so they are referred as full MHD equations. In the first three MHD flow problems considered in Sects. 5.1, 5.2, and 5.3, induced magnetic field is taken into account, and it contributes to the equations as induced magnetic field components (Sects. 5.1 and 5.2), magnetic potential and current density (Sect. 5.3). In Sects. 5.2 and 5.3 magnetic induction equations are accompanied with the energy equation, and Sect. 5.4 contains simply inductionless Buoyancy driven MHD flow. Thus, we present the DRBEM solution of full MHD equations, unidirectional (pipe-axis direction, Sects. 5.5 and 5.6) MHD flow equations, and inductionless MHD flow equations, with or without heat transfer in enclosures.

The first problem considered is the unidirectional unsteady MHD duct flow in which both Re and R_m are not small and multiply the time derivatives of the velocity and induced magnetic field, respectively (Eqs. (1.57)). The equations cannot be decoupled due to the time derivative terms and must be solved as a whole using the DRBEM. The FDM is made use of for the time integration, and the time variations of the velocity and the induced magnetic field are presented (Sect. 5.1).

M. Tezer-Sezgin, C. Bozkaya, *Boundary Element Method for Magnetohydrodynamic Flow*, Surveys and Tutorials in the Applied Mathematical Sciences 14,
https://doi.org/10.1007/978-3-031-58353-7_5

Section 5.2 considers the problem of MHD flow and heat transfer in cavities. Full MHD equations together with energy equation, Eqs. (1.33), in which the momentum equations include Buoyancy force as well as the Lorentz force are solved using the DRBEM. The variations of the solution with the problem parameters as Re, R_m, Pr, Ra, and Ha are discussed and simulated as streamlines, isolines, and contours of vorticity, contours of induced magnetic field components in the lid-driven cavity problem.

In Sect. 5.3, a different form of full MHD equations with heat transfer (Eqs. (1.42)) is solved numerically using the DRBEM and simulated for the MHD flow in the lid-driven cavity. Induced magnetic field is taken into account, and it contributes to the equations as magnetic potential and current density. Stream function, vorticity, magnetic potential, current density, and temperature are obtained for several values of problem parameters.

If the magnetic Reynolds number is small, $R_m << 1$, the magnetic field induced by currents in the fluid is negligible compared to the externally applied magnetic field. In this situation, the flow no longer affects the magnetic field. Section 5.4 presents Buoyancy driven MHD flow described by the flow and temperature equations (1.46) that is an inductionless MHD convection flow.

In Sect. 5.5, the steady, viscous flow of an incompressible, electrically conducting fluid is considered in a channel with electrically conducting walls under an oblique magnetic field applied in plane (perpendicular to the flow direction). The variation of the velocity is assumed to be in the channel axis direction as having one component (z-component) only in Eqs. (1.52) that are written in terms of this velocity component in the flow direction (z-axis) and the electric potential. Solution is depicted for increasing values of Hartmann number and for the most general form of electric potential boundary conditions.

As a last problem in this chapter, MHD flow in terms of the velocity and electric potential is considered in piping systems where the duct walls are electrically conducting. The inductionless MHD flow in channels under the effect of magnetic field applied in the pipe-axis direction, Eqs. (1.54), is solved in Sect. 5.6 in terms of stream function–vorticity–electric potential and pipe-axis velocity. The solution is simulated for several values of Hartmann and Stuart numbers.

5.1 DRBEM-FDM Solution of Time-Dependent MHD Duct Flow

The time-dependent MHD duct flow problem given in Chap. 1 (Sect. 1.2.8) with the nondimensional equations (1.57) is considered in a square duct ($\Omega = \{(x, y) |\, |x| \leq 1,\ |y| \leq 1\}$) subject to a horizontally applied magnetic field, that is, $Ha_x = Ha$ and $Ha_y = 0$ in (1.57). Thus, the system of Eqs. (1.57) for the velocity V and induced magnetic field B are reduced to

$$\begin{aligned} \nabla^2 V + Ha\frac{\partial B}{\partial x} &= -1 + Re\frac{\partial V}{\partial t} \\ \nabla^2 B + Ha\frac{\partial V}{\partial x} &= R_m\frac{\partial B}{\partial t}\,. \end{aligned} \tag{5.1}$$

The boundary and initial conditions are (no-slip velocity and insulated walls) are accompanied to Eq. (5.1)

$$\begin{aligned} V(x, y, t) = B(x, y, t) = 0, &\quad (x, y) \in \partial\Omega \\ V(x, y, 0) = B(x, y, 0) = 0, &\quad (x, y) \in \Omega\,. \end{aligned} \tag{5.2}$$

The unsteady MHD flow equations are studied widely with some numerical methods used in CFD (computational fluid dynamics) for the case of $Re = R_m = 1$. The unsteady MHD flow in a duct having insulated walls has been solved by Tezer-Sezgin and Gürbüz [70] using radial basis function approximation. The meshless local boundary integral equation method has been presented for transient MHD flow through rectangular and circular pipes in Dehghan and Mirzai [71]. The study conducted by Bozkaya and Tezer-Sezgin [72] using time-domain BEM and a combination of the DRBEM and Differential Quadrature Method (DQM) by Bozkaya and Tezer-Sezgin [73] have been given for unsteady MHD flow in a rectangular duct having insulated walls.

Equations (5.1) in their original coupled form are transformed into equivalent boundary integral equations by using the dual reciprocity BEM in which the fundamental solution of Laplace equation is employed. The convective and time derivative terms are approximated by using the coordinate matrix as explained in Sects. 2.2.3 and 2.2.4, respectively. Thus, the analogous of DRBEM equations (2.66) for the present problem becomes

$$\begin{aligned} \mathbf{HV} - \mathbf{GV_q} &= \mathbf{C}(\{\mathbf{-1}\} - Ha\frac{\partial \mathbf{F}}{\partial x}\mathbf{F^{-1}B} + Re\frac{\partial \mathbf{V}}{\partial t}) \\ \mathbf{HB} - \mathbf{GB_q} &= \mathbf{C}(-Ha\frac{\partial \mathbf{F}}{\partial x}\mathbf{F^{-1}V} + R_m\frac{\partial \mathbf{B}}{\partial t}) \end{aligned} \tag{5.3}$$

in $\mathbf{V}$, $\mathbf{B}$, $\mathbf{V_q} = \frac{\partial \mathbf{V}}{\partial n}$, $\mathbf{B_q} = \frac{\partial \mathbf{B}}{\partial n}$, where $\mathbf{C} = (\mathbf{H\hat{U}} - \mathbf{G\hat{Q}})\mathbf{F^{-1}}$ and $\{\mathbf{-1}\}$ denotes the vector with components -1. The unconditionally stable backward difference integration scheme defined by

$$\frac{\partial u}{\partial t}\Big|_{n+1} = \frac{u^{n+1} - u^n}{\Delta t}$$

is used for the time integration where n indicates the time level. Thus, the time discretized form of DRBEM system of equations for the velocity and induced magnetic field becomes

$$\begin{aligned}(\mathbf{H}-\frac{Re}{\Delta t}\mathbf{C})\mathbf{V}^{n+1}+Ha\mathbf{C}\frac{\partial\mathbf{F}}{\partial x}\mathbf{F}^{-1}\mathbf{B}^{n+1}&=\mathbf{G}\mathbf{V}_{\mathbf{q}}^{n+1}-\mathbf{C}\{\mathbf{1}\}-\frac{Re}{\Delta t}\mathbf{C}\mathbf{V}^{n}\\(\mathbf{H}-\frac{R_m}{\Delta t}\mathbf{C})\mathbf{B}^{n+1}+Ha\mathbf{C}\frac{\partial\mathbf{F}}{\partial x}\mathbf{F}^{-1}\mathbf{V}^{n+1}&=\mathbf{G}\mathbf{B}_{\mathbf{q}}^{n+1}-\frac{R_m}{\Delta t}\mathbf{C}\mathbf{B}^{n}\end{aligned}\quad , \tag{5.4}$$

which can be written in matrix–vector form as follows:

$$\begin{bmatrix}\mathbf{H}-\frac{Re}{\Delta t}\mathbf{C} & Ha\mathbf{C}\frac{\partial\mathbf{F}}{\partial x}\mathbf{F}^{-1}\\ Ha\mathbf{C}\frac{\partial\mathbf{F}}{\partial x}\mathbf{F}^{-1} & \mathbf{H}-\frac{R_m}{\Delta t}\mathbf{C}\end{bmatrix}\begin{bmatrix}\mathbf{V}\\ \mathbf{B}\end{bmatrix}^{n+1}=\begin{bmatrix}\mathbf{G} & \mathbf{0}\\ \mathbf{0} & \mathbf{G}\end{bmatrix}\begin{bmatrix}\mathbf{V}_{\mathbf{q}}\\ \mathbf{B}_{\mathbf{q}}\end{bmatrix}^{n+1}$$

$$-\begin{bmatrix}\frac{Re}{\Delta t}\mathbf{C}\\ \frac{R_m}{\Delta t}\mathbf{C}\end{bmatrix}\begin{bmatrix}\mathbf{V}\\ \mathbf{B}\end{bmatrix}^{n}-\begin{bmatrix}\mathbf{C}\{\mathbf{1}\}\\ \mathbf{0}\end{bmatrix}. \tag{5.5}$$

After the insertion of boundary conditions in system (5.5), the solution is obtained iteratively for the transient time levels starting with the zero initial velocity and initial induced magnetic field.

In the numerical simulations, computations are performed for various values of dimensionless parameters Reynolds number (Re), magnetic Reynolds number (R_m), and Hartman number (Ha) in order to analyze their effects on the velocity and induced magnetic field. First, the time variations of the velocity and induced magnetic field along the centerline $y = 0$ at transient levels are displayed in Fig. 5.1 for $Ha = 5, 20, 50, 100$, $Re = 1$, $R_m = 1$. It is noticed that the steady

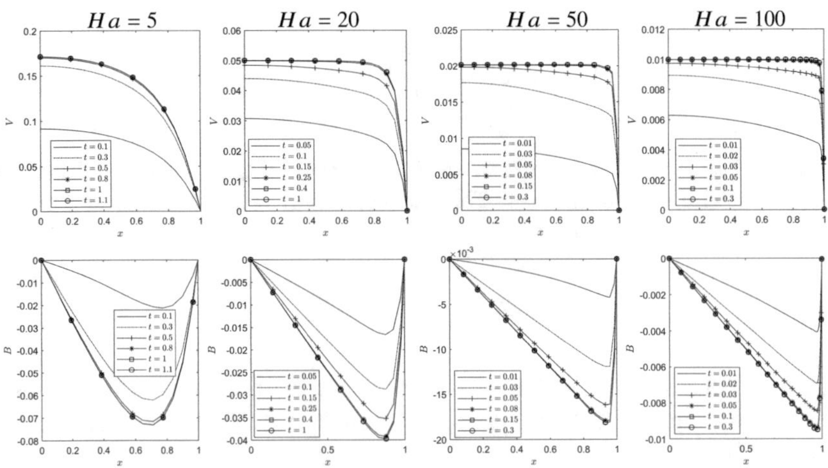

Fig. 5.1 Time variations of the velocity and induced magnetic field along the centerline $y = 0$, $0 \leq x \leq 1$ for $Ha = 5, 20, 50, 100$, $Re = 1$, $R_m = 1$

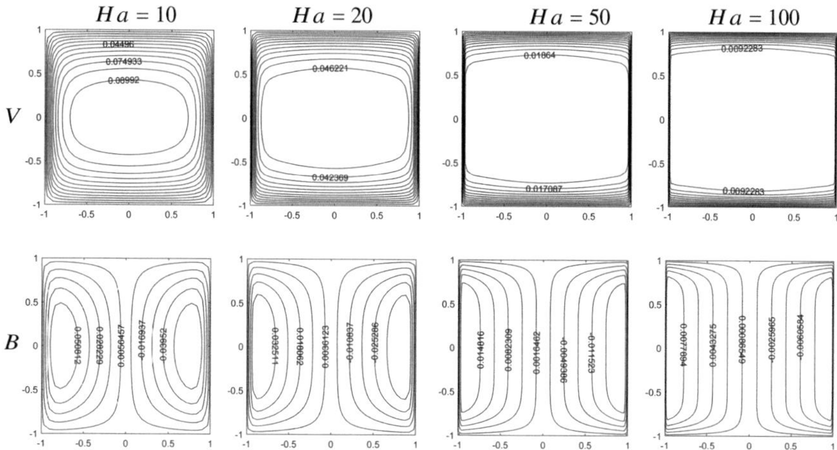

Fig. 5.2 Effect of Ha on equi-velocity and current lines at steady state ($t = 1$): $Re = 1$, $R_m = 1$

states for the velocity and induced magnetic field are reached in earlier time levels for higher values of Ha. The steady-state values of V and B are obtained at $t \approx 0.5, 0.25, 0.08, 0.05$ when $Ha = 5, 20, 50, 100$, respectively. The well-known characteristics of the MHD duct flow is observed. That is, both the flow and the induced magnetic field display a flattening tendency and form boundary layers close to the walls of the duct as Ha increases.

The equi-velocity and current lines are shown in Fig. 5.2 for values of $Ha(= 10, 20, 50, 100)$ at fix $Re = R_m = 1$ in order to see the effect of the strength of the externally applied magnetic field. The formation of the Hartmann layers of thickness $1/Ha$ along the vertical walls and the side layers of thickness $1/\sqrt{Ha}$ along the horizontal walls is more pronounced as Ha increases, while the values of velocity and the induced magnetic field are decreasing.

The variations of the velocity and induced magnetic field along the centerline $y = 0$ are displayed in Fig. 5.3 for various combinations of Re and R_m at time levels $t = 1, 5, 50$. The computations are carried out at fix values of $Ha = 20$, $R_m = 1$ and at $Ha = 20$, $Re = 1$ to analyze the effects of Re and R_m, respectively. The velocity magnitude of the fluid decreases with an increase in Re at all the time levels as it is known from the analysis of the ordinary fluid flows. However, this decrease slows down when the time increases. Similarly, an increase in both Re and R_m decreases the induced magnetic field in magnitude, also losing its decrease strength for increasing time levels. On the other hand, the velocity increases as R_m increases at transient time levels $t = 1, 5$. From the definition of the magnetic Reynolds number, it can be seen that an increase in R_m occurs when either the magnetic permeability or the electrical conductivity of the fluid is increased, which naturally increases the velocity of the fluid. However, R_m has no effect on the values of velocity or the induced magnetic field at further time levels (i.e., $t = 50$).

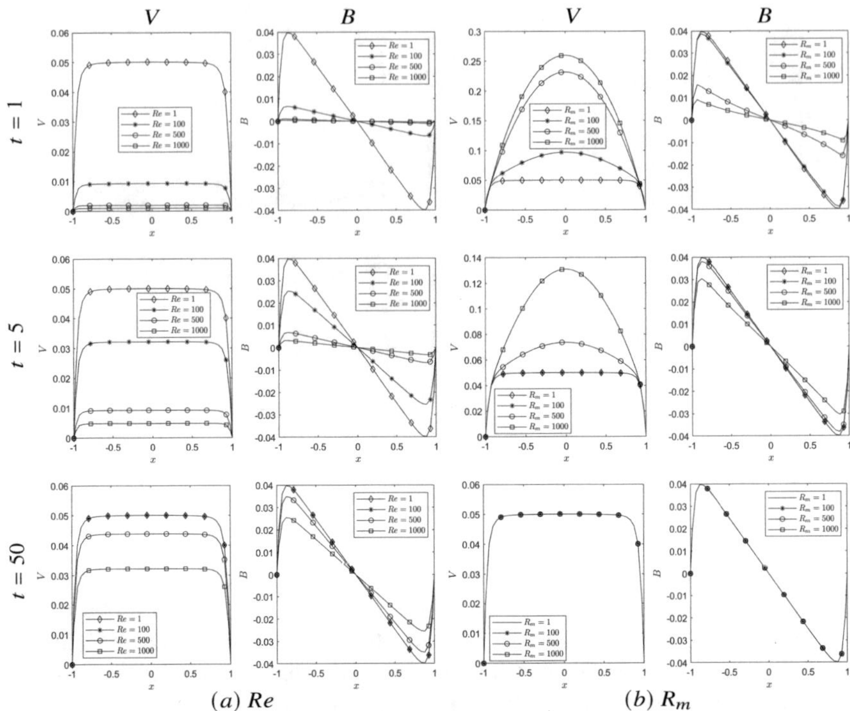

Fig. 5.3 Effects of (**a**) Re when $Ha = 20$, $R_m = 1$ and (**b**) R_m when $Ha = 20$, $Re = 1$ on the distribution of the velocity and induced magnetic field along the centerline $y = 0$, $-1 \leq x \leq 1$ at time levels $t = 1$(top), $t = 5$(middle) and $t = 50$(bottom)

The transient behavior of the MHD duct flow under a time-varied external magnetic field is given in Ebren Kaya and Tezer-Sezgin [74] using the DRBEM. The effects of Re, R_m and the time-varied function in the magnetic field are simulated in a cavity in terms of the velocity and the induced magnetic field lines.

5.2 MHD Flow and Heat Transfer with Magnetic Induction in Lid-Driven Cavity

This section is devoted to the DRBEM solution of the full MHD equations in terms of stream function ψ, vorticity w, temperature T, and the induced magnetic field components B_x, B_y, which are given in dimensionless equation (1.33) in Chap. 1 (Sect. 1.2.4) as follows:

$$
\begin{aligned}
\nabla^2\psi &= -w \\
\frac{1}{Re}\nabla^2 w &= \frac{\partial w}{\partial t} + u\frac{\partial w}{\partial x} + v\frac{\partial w}{\partial y} - \frac{Ra}{PrRe^2}\frac{\partial T}{\partial x} \\
&\quad - \frac{Ha^2}{ReR_m}[B_x\frac{\partial}{\partial x}(\frac{\partial B_y}{\partial x} - \frac{\partial B_x}{\partial y}) + B_y\frac{\partial}{\partial y}(\frac{\partial B_y}{\partial x} - \frac{\partial B_x}{\partial y})] \\
\frac{1}{R_m}\nabla^2 B_x &= \frac{\partial B_x}{\partial t} + u\frac{\partial B_x}{\partial x} + v\frac{\partial B_x}{\partial y} - B_x\frac{\partial u}{\partial x} - B_y\frac{\partial u}{\partial y} \\
\frac{1}{R_m}\nabla^2 B_y &= \frac{\partial B_y}{\partial t} + u\frac{\partial B_y}{\partial x} + v\frac{\partial B_y}{\partial y} - B_x\frac{\partial v}{\partial x} - B_y\frac{\partial v}{\partial y} \\
\frac{1}{PrRe}\nabla^2 T &= \frac{\partial T}{\partial t} + u\frac{\partial T}{\partial x} + v\frac{\partial T}{\partial y}.
\end{aligned}
\tag{5.6}
$$

The nondimensional physical parameters are the Prandtl (Pr), Reynolds (Re), magnetic Reynolds (R_m), Rayleigh (Ra), and Hartmann (Ha) numbers.

Equations (5.6) are discretized by using the DRBEM in space with the fundamental solution of Laplace equation, $u^* = 1/2\pi \ln(1/r)$, and backward-Euler finite difference scheme in time as explained in Sect. 2.2.4. Thus, the equations in (5.6) are weighted by u^*, and the application of Green's second identity results in [3]

$$
c_i S_i + \int_\Gamma (q^* S - u^* \frac{\partial S}{\partial n}) d\Gamma = -\int_\Omega b_S u^* d\Omega, \tag{5.7}
$$

where S stands for each unknown ψ, w, B_x, B_y, T. Here, $q^* = \partial u^*/\partial n$, Γ is the boundary of the domain Ω, and the constant $c_i = \theta_i/2\pi$ with the internal angle θ_i at the source point i. All the terms on the right-hand side of Eq. (5.6) denoted by b_S are approximated by a set of linear radial basis functions $f_j (= 1 + r_j)$ linked with the particular solutions $\hat{u}_j$ of $\nabla^2 \hat{u}_j = f_j$, $j = 1, \cdots, N + L$, N and L being the number of boundary and interior nodes, respectively [3]. When Green's identity is applied to the right-hand side as well, and the boundary is discretized with constant elements, the matrix–vector form of Eq. (5.7) can be expressed as

$$
\mathbf{H}S - \mathbf{G}\frac{\partial S}{\partial n} = (\mathbf{H}\hat{\mathbf{U}} - \mathbf{G}\hat{\mathbf{Q}})\mathbf{F}^{-1} b_S, \tag{5.8}
$$

where the components of matrices $\mathbf{H}$ and $\mathbf{G}$ are calculated by integrating q^* and u^*, respectively, over each boundary element as given in Eq. (2.23). The matrices $\hat{\mathbf{U}}$, $\hat{\mathbf{Q}}$, and $\mathbf{F}$ take the vectors $\hat{u}_j$, $\hat{q}_j$, f_j as columns, respectively, and S, b_S are of size $(N + L) \times 1$.

When the backward finite difference is applied to approximate the time derivatives in Eq. (5.8), and right-hand sides are substituted by corresponding b_S at $N+L$

points, the DRBEM system of algebraic equations for Eqs. (5.6) takes the form

$$
\begin{aligned}
&\mathbf{H}\psi^{n+1} - \mathbf{G}\psi_q^{n+1} = -\mathbf{C}w^n \\
&(\mathbf{H} - \frac{Re}{\Delta t}\mathbf{C} - Re\mathbf{CK})w^{n+1} - \mathbf{G}w_q^{n+1} = -\frac{Re}{\Delta t}\mathbf{C}w^n - \frac{Ra}{Pr\,Re}\mathbf{CD_x}T^{n+1} \\
&\qquad\qquad - \frac{Ha^2}{R_m}\mathbf{C}([\mathbf{B_x}]_d^{n+1}\mathbf{D_x}R + [\mathbf{B_y}]_d^{n+1}\mathbf{D_y}R) \\
&(\mathbf{H} - \frac{R_m}{\Delta t}\mathbf{C} - R_m\mathbf{CK} + R_m\mathbf{CD_x}[\mathbf{u}]_d^{n+1})B_x^{n+1} - \mathbf{G}B_{xq}^{n+1} = -\frac{R_m}{\Delta t}\mathbf{C}B_x^n \\
&\qquad\qquad - R_m\mathbf{C}[\mathbf{B_y}]_d^n\mathbf{D_y}u^{n+1} \\
&(\mathbf{H} - \frac{R_m}{\Delta t}\mathbf{C} - R_m\mathbf{CK} + R_m\mathbf{CD_y}[\mathbf{v}]_d^{n+1})B_y^{n+1} - \mathbf{G}B_{yq}^{n+1} = -\frac{R_m}{\Delta t}\mathbf{C}B_y^n \\
&\qquad\qquad - R_m\mathbf{C}[\mathbf{B_x}]_d^n\mathbf{D_x}v^{n+1} \\
&(\mathbf{H} - \frac{Pr\,Re}{\Delta t}\mathbf{C} - Pr\,Re\mathbf{CK})T^{n+1} - \mathbf{G}T_q^{n+1} = \frac{Pr\,Re}{\Delta t}\mathbf{C}T^n ,
\end{aligned}
\tag{5.9}
$$

where $\mathbf{C} = (\mathbf{H}\hat{\mathbf{U}} - \mathbf{G}\hat{\mathbf{Q}})\mathbf{F}^{-1}$, $\mathbf{K} = [\mathbf{u}]_d^{n+1}\mathbf{D_x} + [\mathbf{v}]_d^{n+1}\mathbf{D_y}$, $u^{n+1} = \mathbf{D_y}\psi^{n+1}$, $v^{n+1} = -\mathbf{D_x}\psi^{n+1}$, $\mathbf{D_x} = \frac{\partial \mathbf{F}}{\partial x}\mathbf{F}^{-1}$, $\mathbf{D_y} = \frac{\partial \mathbf{F}}{\partial y}\mathbf{F}^{-1}$, and $R = \mathbf{D_x}B_y^{n+1} - \mathbf{D_y}B_x^{n+1}$. The unknown vectors ψ, w, B_x, B_y, and T are of size $(N+L) \times 1$. The matrices $[\mathbf{B_x}]_d^{n+1}$, $[\mathbf{B_y}]_d^{n+1}$, and $[\mathbf{u}]_d^{n+1}$ and $[\mathbf{v}]_d^{n+1}$ containing u^{n+1} and v^{n+1} as entries, respectively, are diagonal matrices of size $(N + L)$. This system of coupled equations is solved iteratively for increasing time levels with zero initial estimates of w, B_x, B_y, and T except on the boundary. In each time level, the space derivatives of each unknown denoted by the vector S, and the missing boundary conditions of the vorticity are obtained by using the coordinate matrix $\mathbf{F}$ as

$$
\frac{\partial S}{\partial x} = \mathbf{D_x}S, \ \frac{\partial S}{\partial y} = \mathbf{D_y}S, \ w = -(\frac{\partial^2 \mathbf{F}}{\partial x^2} + \frac{\partial^2 \mathbf{F}}{\partial y^2})\mathbf{F}^{-1}\psi .
$$

The solution process continues until the criterion

$$
\sum_{k=1}^{m} \frac{\left\| S_k^{n+1} - S_k^n \right\|_\infty}{\left\| S_k^{n+1} \right\|_\infty} < \epsilon \tag{5.10}
$$

is reached for a time level n where m is the number of unknowns and ϵ is a preassigned tolerance.

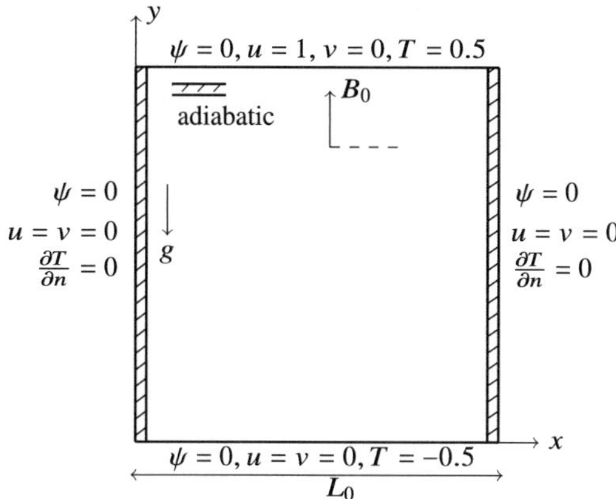

Fig. 5.4 Physical configuration of the problem and the boundary conditions

The numerical simulations are performed in a square lid-driven cavity beneath a vertically applied uniform magnetic field ($\mathbf{B} = (0, B_0, 0)$) in order to investigate the effects of the problem parameters Re, R_m, Ha, Ra at a fixed $Pr = 0.1$ on the flow, temperature, and the induced magnetic field. The problem configuration and the boundary conditions are shown in Fig. 5.4. The walls of the cavity with side length $L_0 = 1$ are discretized by using maximum $N = 200$ constant boundary elements and $L = 2500$ interior nodes for the cases when the highest values of the problem parameters are considered. The preassigned tolerance ϵ is set to be 10^{-4} in the stopping criteria (5.10) of the iterative schemes; and the solutions in regard to this criteria are referred as the steady-state solutions.

As Re increases (Fig. 5.5), the central vortex in the flow shifts through the center of the cavity. Secondary eddies are formed near the bottom corner, the right one being larger due to the movement of the upper lid to the right. Vorticity shows concentration in terms of boundary layers through upper right corner and on the top moving lid. The effect of Re is quite slight on the induced magnetic field. For $Re \geq 400$ isotherms form strong temperature gradients clustered at the left top and right bottom corners.

The increase of R_m influences the induced magnetic field showing circulation at the center of the cavity due to the dominance of convection terms in the induction equations, as can be seen in Fig. 5.6.

Figure 5.7 displays the effect of the Rayleigh number Ra on the flow and the induced magnetic field. The increase in Ra slightly alters the isotherms due to the dominance of buoyancy force. The strong effect is observed on the streamlines when $Ra = 10^4$, dividing the cavity into two counter flows and even forming the third loop in front of the bottom wall when it reaches to $Ra = 10^5$. Induced magnetic field is completely directed from the bottom to the top wall as Ra increases.

As Ha increases (Fig. 5.8), the flow is retarded, that is, the magnitude of streamlines drops. The flow tends to be separated into two cells near the left and right side walls forming also boundary layers for $Ha \geq 100$. The transfer of the heat from the hot wall to the cold wall turns to be directly in between the adiabatic

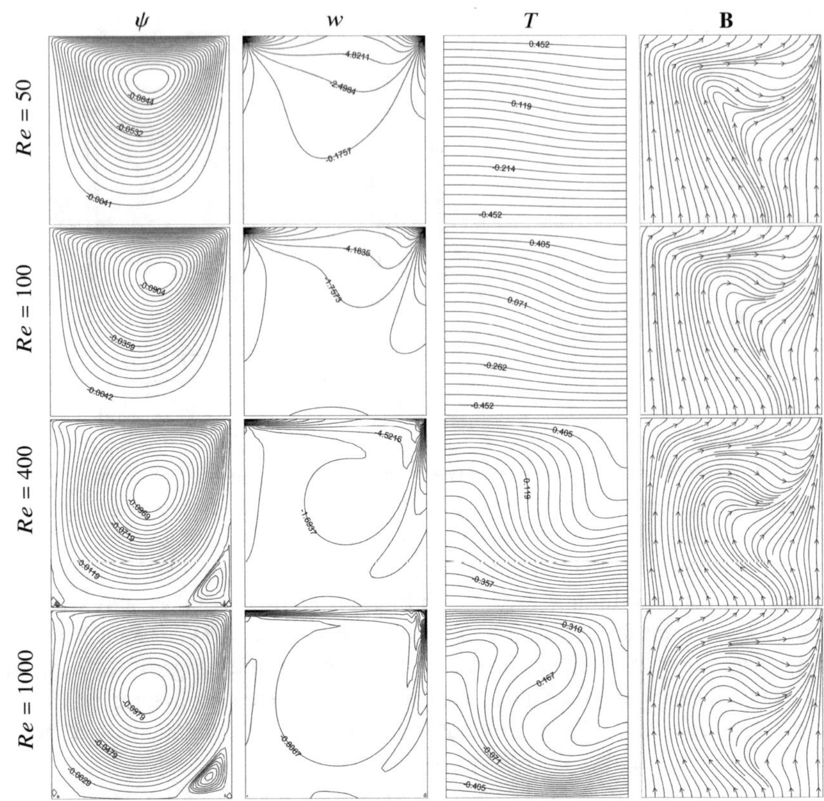

Fig. 5.5 Effects of Re when $Pr = 0.1$, $Ha = 10$, $Ra = 100$, $R_m = 100$, $\Delta t = 0.25$

left and right side walls. An increase in Ha has similar effect on induced current as if Ra increases.

The same lid-driven MHD flow problem in terms of full MHD equations has been solved in [75, 76] by using DRBEM. The numerical results obtained here are in very well agreement with the ones in [76] for the same values of problem parameters.

5.3 Buoyancy MHD Flow with Magnetic Potential

In this section, the full MHD flow equations given in Sect. 5.2 are considered with the magnetic potential A and the current density j instead of the components of the induced magnetic field B_x and B_y. The derivation of these equations and their nondimensional form are given in Chap. 1 and described with Eq. (1.42) as

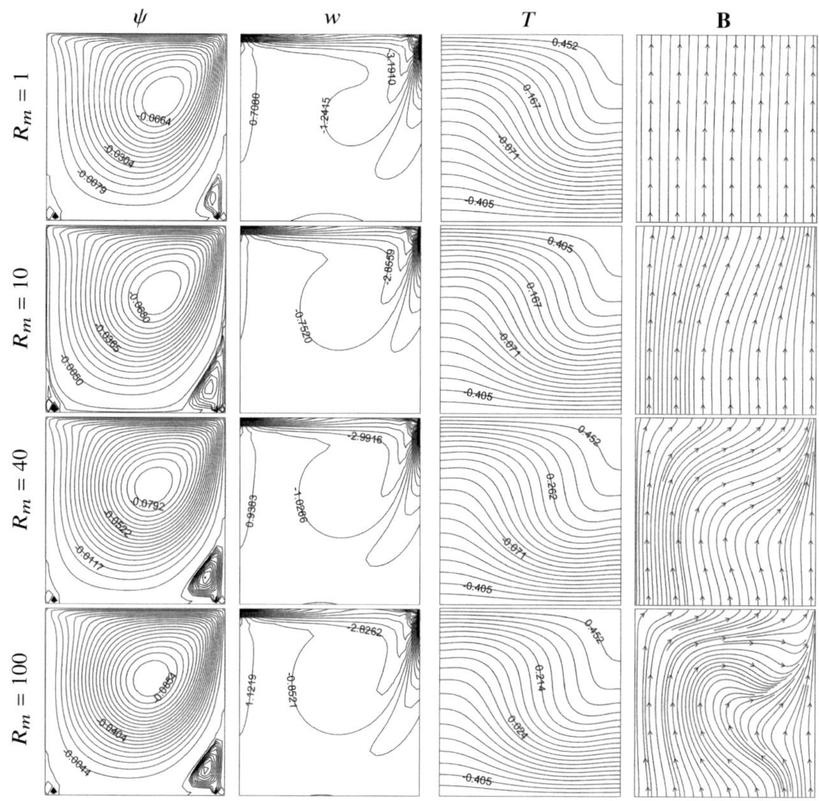

Fig. 5.6 Effect of R_m when $Pr = 0.1$, $Ha = 10$, $Ra = 10^3$, $Re = 400$, $\Delta t = 0.25$

$$
\begin{aligned}
\nabla^2 \psi &= -w \\
\frac{1}{Re}\nabla^2 w &= \frac{\partial w}{\partial t} + u\frac{\partial w}{\partial x} + v\frac{\partial w}{\partial y} - \frac{Ra}{Pr\,Re^2}\frac{\partial T}{\partial x} - \frac{Ha^2}{Re}(B_x\frac{\partial j}{\partial x} + B_y\frac{\partial j}{\partial y}) \\
\nabla^2 A &= -R_m\, j \\
\frac{1}{R_m}\nabla^2 j &= \frac{\partial j}{\partial t} + u\frac{\partial j}{\partial x} + v\frac{\partial j}{\partial y} - \frac{1}{R_m}(B_x\frac{\partial w}{\partial x} + B_y\frac{\partial w}{\partial y}) \\
&\quad - \frac{2}{R_m}\left[\frac{\partial B_x}{\partial x}(\frac{\partial v}{\partial x} + \frac{\partial u}{\partial y}) + \frac{\partial v}{\partial y}(\frac{\partial B_x}{\partial y} + \frac{\partial B_y}{\partial x})\right] \\
\frac{1}{Pr\,Re}\nabla^2 T &= \frac{\partial T}{\partial t} + u\frac{\partial T}{\partial x} + v\frac{\partial T}{\partial y}\,,
\end{aligned}
\tag{5.11}
$$

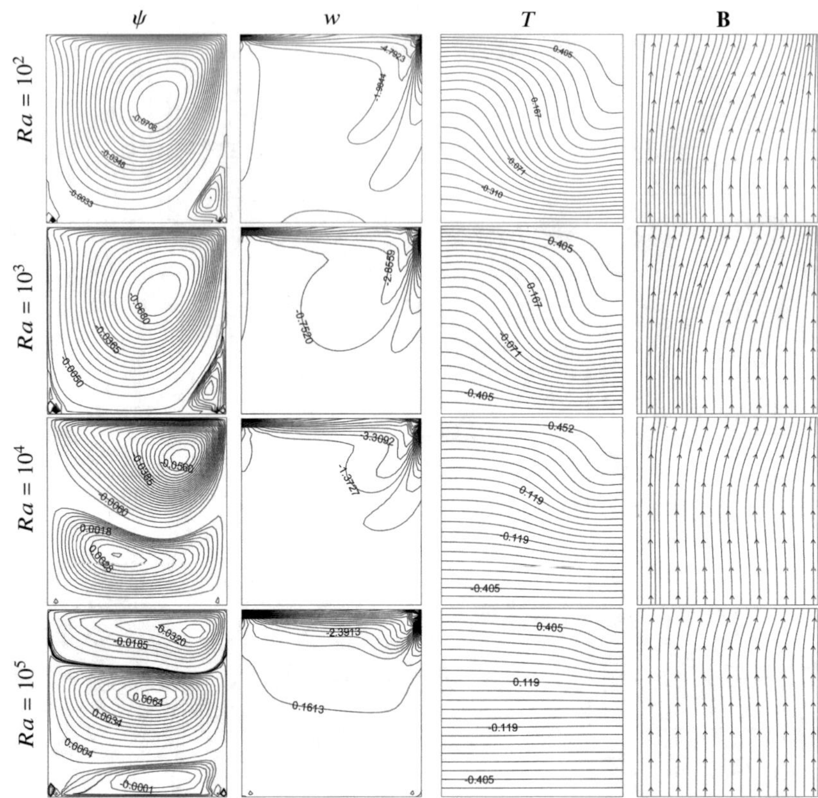

Fig. 5.7 Effect of Ra when $Pr = 0.1$, $Ha = R_m = 10$, $Re = 400$, $\Delta t = 0.25$

where the magnetic potential A is defined as $B_x = \partial A/\partial y$, $B_y = -\partial A/\partial x$ replacing B_x, B_y in Eqs. (5.11). Also, the current density j is given as $j = \frac{1}{R_m}(\partial B_y/\partial x - \partial B_x/\partial y)$, where j is the z-component of electric current density $\mathbf{J} = (0, 0, j)$ from Eq. 1.36. The application of the DRBEM and backward-Euler finite difference scheme discretize these equations as explained in the previous Sect. 5.2. Thus, the resulting discretized DRBEM equations in matrix–vector form become

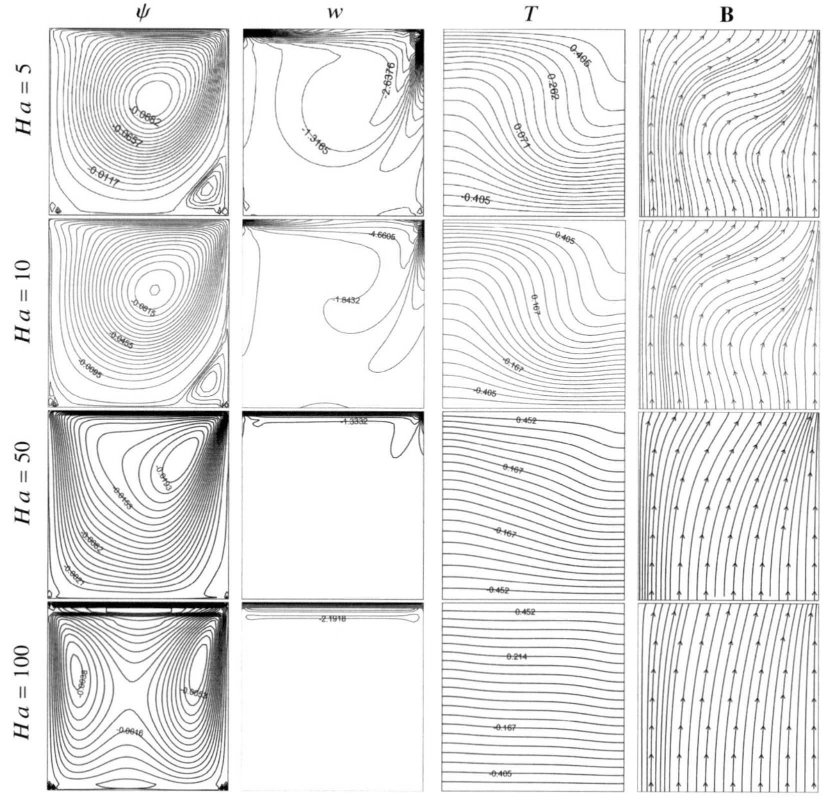

Fig. 5.8 Effects of $Ha = 5, 10, 50$ ($\Delta t = 0.2$) and $Ha = 100$ ($\Delta t = 0.1$) when $Pr = 0.1$, $Ra = 10^3$, $R_m = 40$, $Re = 400$

$$\mathbf{H}\psi^{n+1} - \mathbf{G}\psi_q^{n+1} = -\mathbf{C}w^n$$

$$\mathbf{H}A^{n+1} - \mathbf{G}A_q^{n+1} = -R_m\mathbf{C}j^n$$

$$(\mathbf{H} - \frac{Re}{\Delta t}\mathbf{C} - Re\mathbf{CK})w^{n+1} - \mathbf{G}w_q^{n+1} = -\frac{Re}{\Delta t}\mathbf{C}w^n - \frac{Ra}{Pr\,Re}\mathbf{C}\mathbf{D_x}T^{n+1}$$

$$-Ha^2\mathbf{C}([\mathbf{B_x}]_d\mathbf{D_x} + [\mathbf{B_y}]_d\mathbf{D_y}R)j^{n+1}$$

$$(\mathbf{H} - \frac{R_m}{\Delta t}\mathbf{C} - R_m\mathbf{CK})j^{n+1} - \mathbf{G}j_q^{n+1} = -\frac{R_m}{\Delta t}\mathbf{C}j^n - \mathbf{C}([\mathbf{B_x}]_d\mathbf{D_x} + [\mathbf{B_y}]_d\mathbf{D_y})w^n$$

$$-2\mathbf{C}(\mathbf{D_x}[\mathbf{B_x}]_d(\mathbf{D_x}v^{n+1} + \mathbf{D_y}u^{n+1}) - \mathbf{D_y}[\mathbf{v}]_d(\mathbf{Dy}B_x^{n+1} + \mathbf{D_x}B_y^{n+1}))$$

$$(\mathbf{H} - \frac{Pr\,Re}{\Delta t}\mathbf{C} - Pr\,Re\mathbf{CK})T^{n+1} - \mathbf{G}T_q^{n+1} = \frac{Pr\,Re}{\Delta t}\mathbf{C}T^n \tag{5.12}$$

in which all the matrices and vectors are as defined in Sect. 5.2. The lid-driven cavity flow problem given in Sect. 5.2 with the same boundary conditions for ψ, T, u, v is also solved by taking $A = -x$ on the boundary. The unknown boundary conditions for the current density j are calculated similar to the vorticity boundary conditions derivation by the formula $j = \frac{1}{R_m}(\mathbf{D_x}B_y - \mathbf{D_y}B_x)$, while the components of induced magnetic field are obtained by $B_x = \mathbf{D_y}A$ and $B_y = -\mathbf{D_x}A$.

The flow, temperature, induced current, and magnetic potential behaviors are simulated by using the same number of constant boundary and interior nodes as in Sect. 5.2 for the corresponding values of the problem parameters Re, R_m, Ra, and Ha.

Figure 5.9 displays that main circulation in the flow pattern is shifted through the center of the cavity as Re increases. Boundary layer is concentrated near the upper right corner when Re is reached to the value 1000. Strong boundary layer formation is also seen in the vorticity and current density around the upper right corner with the effect of moving top wall and increase in Re. Temperature gradient increases through the top and bottom walls pointing to the increase in convection. Magnetic potential A and induced magnetic field B circulations at the center of the cavity are pushed through the upper right corner when Re increases, and almost diminish for further increase of Re.

In Fig. 5.10, we present contour lines of problem variables for increasing values of R_m to depict the behavior changes in the solution. Not much of alterations are seen in the flow, vorticity, and isolines; however, counterclockwise flow cells start to be formed at the bottom corners. Magnetic potential, current density, and induced magnetic field are affected for $R_m \geq 40$ in the sense that new cells near the upper right corner emerge since the effect of diffusion term in these equations is weakened when R_m increases.

Simulations of problem variables for increasing values of Ra are given in Fig. 5.11. Counter rotating cells in the flow become prominent due to the increase in buoyancy force and moving upper lid. The circulation of vorticity is shrunk through the upper lid leaving almost all parts of the cavity stagnant. As Ra increases, buoyancy effect is increasing and causes the isolines to become almost parallel to the bottom cold and upper hot walls. Magnetic potential and induced magnetic field are not much affected with the increase in Ra. The center and top cells of current density that are shown up to $Ra = 10^3$ are unified to one cell in front of the moving lid.

Finally, as Ha increases up to 100, streamlines tend to be symmetrically situated forming one central vortex, escaping from the effect of moving lid as seen from Fig. 5.12. Vorticity and current density contours have similar behavior as Ha increases, which is the clustering through the top and bottom walls leaving stagnant center region. Magnetic potential and induced magnetic field stay the same for $Ha \geq 10$ since R_m is not changed. They obey the direction of externally applied magnetic field when Ha is increased.

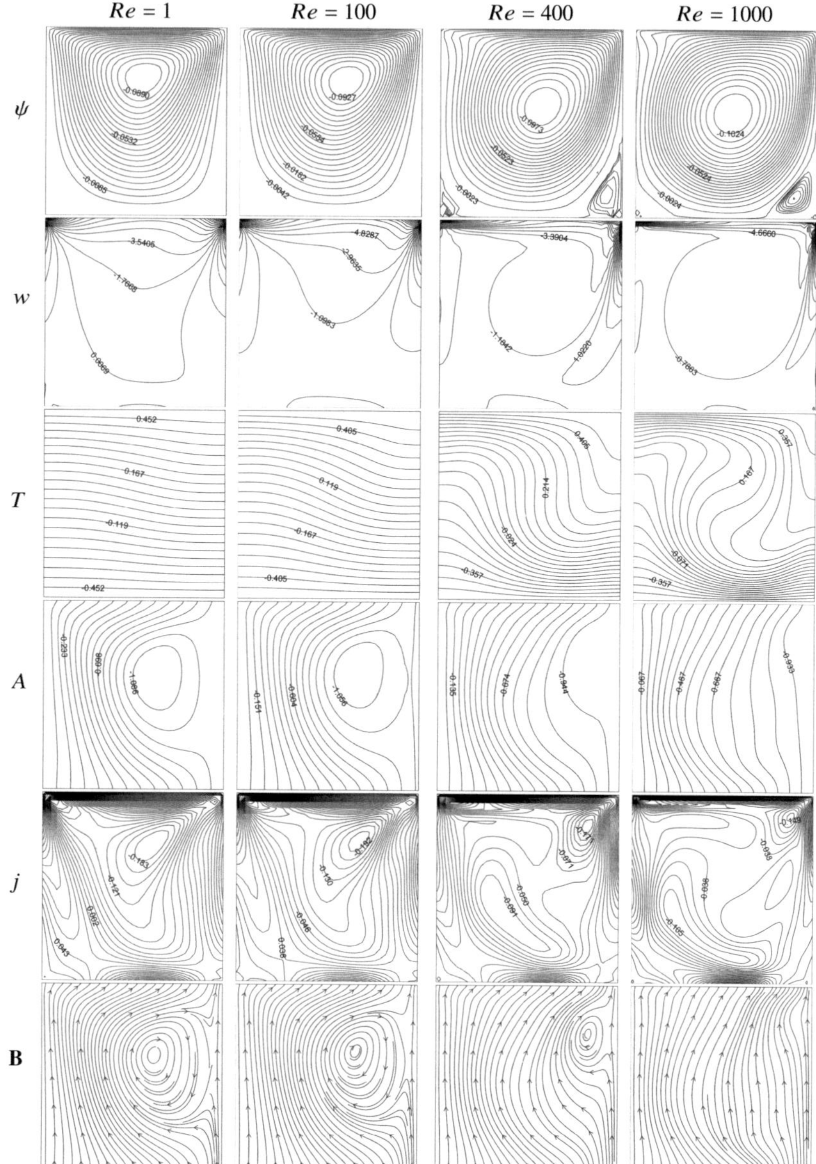

Fig. 5.9 Effects of Re when $Pr = 0.1$, $Ha = 10$, $Ra = 100$, $R_m = 100$, $\Delta t = 0.25$

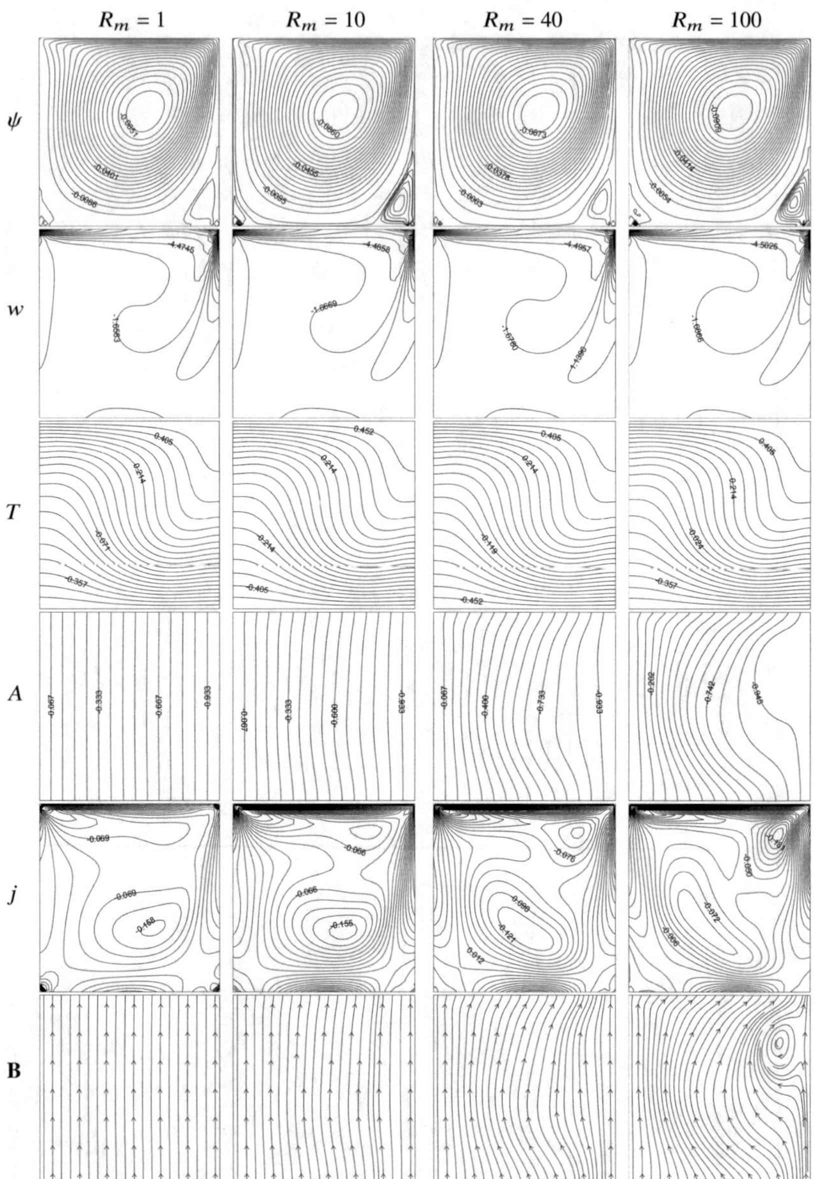

Fig. 5.10 Effects of R_m when $Pr = 0.1$, $Ha = 10$, $Ra = 10^3$, $Re = 400$, $\Delta t = 0.25$

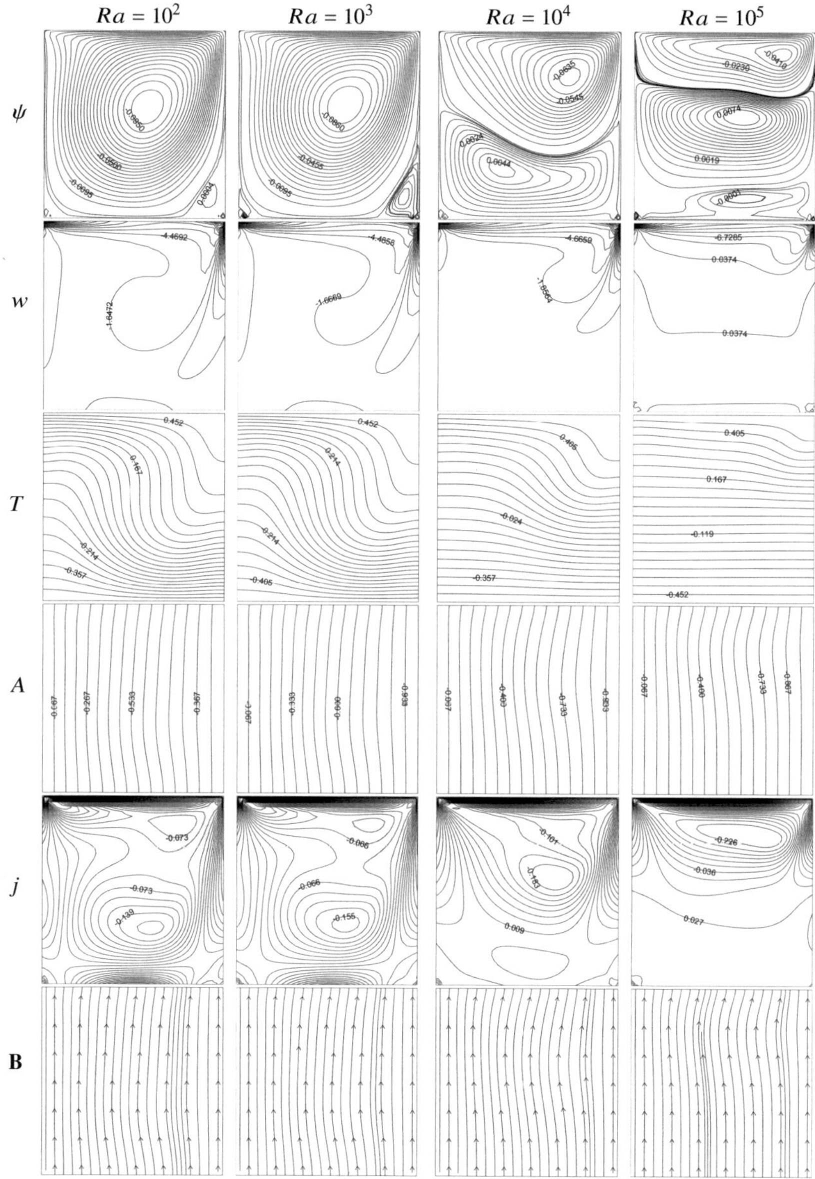

Fig. 5.11 Effects of Ra when $Pr = 0.1$, $Ha = R_m = 10$, $Re = 400$, $\Delta t = 0.25$

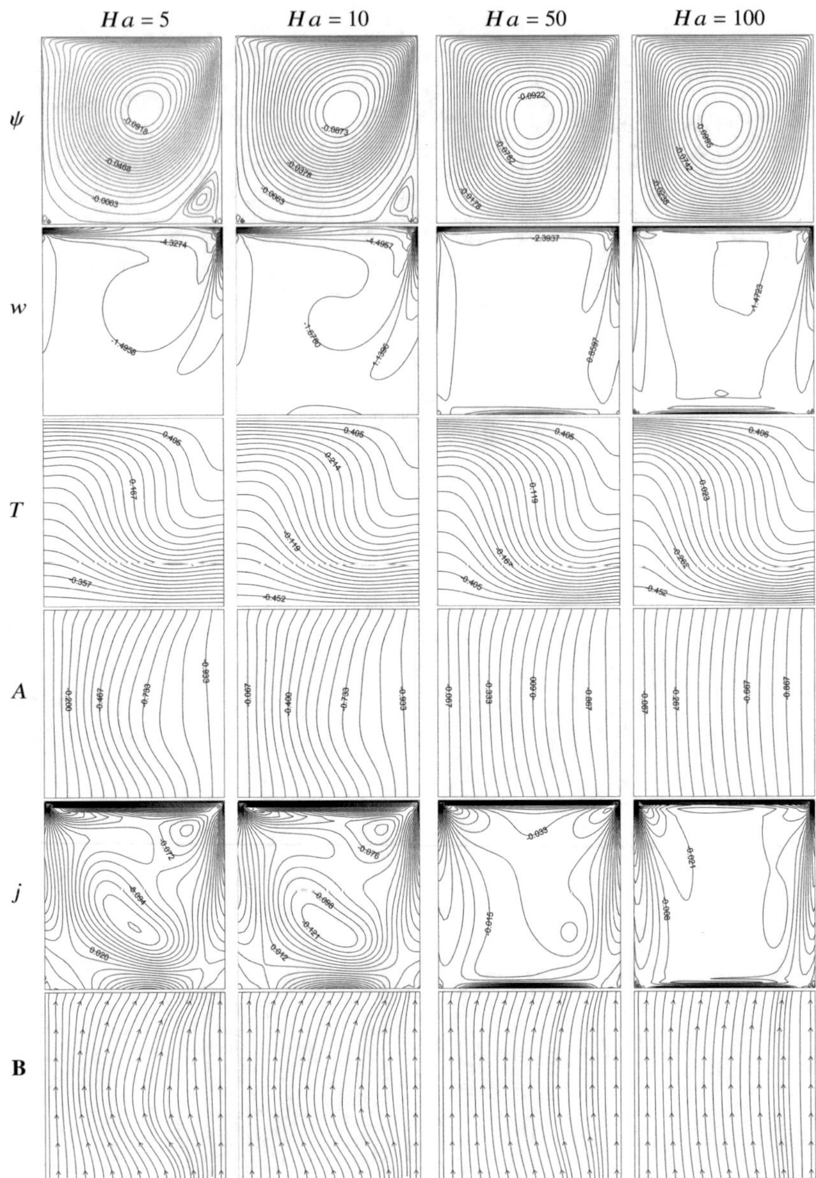

Fig. 5.12 Effects of Ha when $Pr = 0.1$, $Ra = 10^3$, $R_m = 40$, $Re = 400$, $\Delta t = 0.25$

The DRBEM solution of Buoyancy MHD flow with magnetic potential has also been given for staggered double lid-driven cavity and backward facing step flow in [8, 75].

5.4 Inductionless Buoyancy Driven MHD Flow

The magnetic field induced by currents in the fluid can be neglected when R_m is very small as explained in Sect. 1.2.5 of Chap. 1. Thus, the equations involving the induced magnetic field components in the previous two sections are now dropped, and the equations are reduced to Eq. (1.46) in terms of only stream function, vorticity, and temperature

$$\begin{aligned}
\nabla^2\psi &= -w \\
\frac{1}{Re}\nabla^2 w &= \frac{\partial w}{\partial t} + u\frac{\partial w}{\partial x} + v\frac{\partial w}{\partial y} - \frac{Ra}{Pr\,Re^2}\frac{\partial T}{\partial x} - \frac{Ha^2}{Re}\frac{\partial u}{\partial y} \\
\frac{1}{Pr\,Re}\nabla^2 T &= \frac{\partial T}{\partial t} + u\frac{\partial T}{\partial x} + v\frac{\partial T}{\partial y}
\end{aligned} \tag{5.13}$$

when a vertically applied magnetic field ($\mathbf{B} = (0, B_0, 0)$) is considered. The discretized form of these equations after the application of the combined numerical techniques explained in Sect. 5.2 becomes

$$\begin{aligned}
&\mathbf{H}\psi^{n+1} - \mathbf{G}\psi_q^{n+1} = -\mathbf{C}w^n \\
&(\mathbf{H} - \frac{Re}{\Delta t}\mathbf{C} - Re\mathbf{CK})w^{n+1} - \mathbf{G}w_q^{n+1} = -\frac{Re}{\Delta t}\mathbf{C}w^n - \frac{Ra}{Pr\,Re}\mathbf{CD_x}T^{n+1} \\
&\quad -Ha^2\mathbf{CD_y}u^{n+1} \\
&(\mathbf{H} - \frac{Pr\,Re}{\Delta t}\mathbf{C} - Pr\,Re\mathbf{CK})T^{n+1} - \mathbf{G}T_q^{n+1} = \frac{Pr\,Re}{\Delta t}\mathbf{C}T^n\,.
\end{aligned} \tag{5.14}$$

As a test problem again the lid-driven cavity MHD flow described in Sect. 5.2 is considered under the effect of a vertically applied uniform magnetic field. The boundary conditions for the stream function and temperature are taken as given in Fig. 5.4. It can be seen from Figs. 5.13 to 5.14 that the flow and temperature show similar behaviors as observed in Figs. 5.7 and 5.8 of Sect. 5.2, respectively, under the effect of varying Re and Ha, and varying Ra. Flow is almost stagnant at the center of the cavity between the loops near the side walls at $Re = 400$, $Ha = 100$

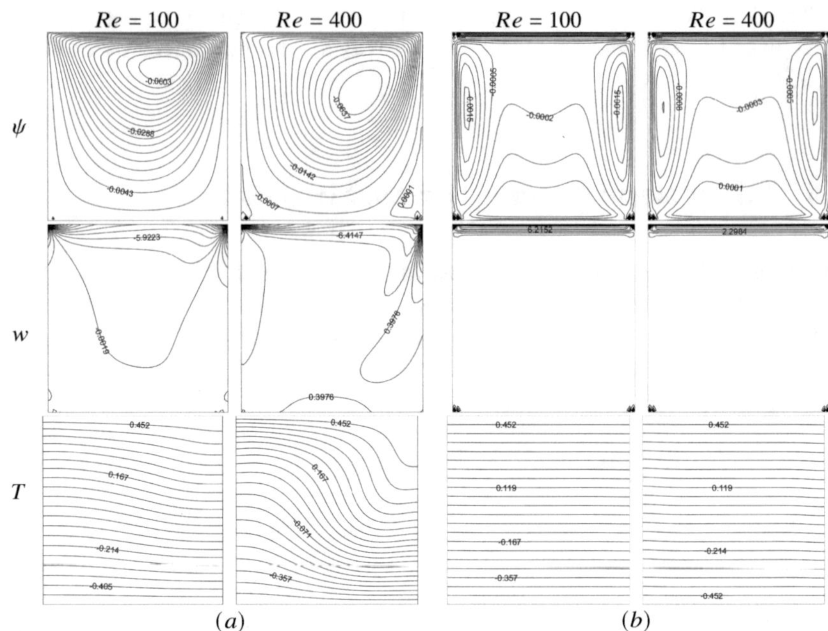

Fig. 5.13 Effect of Ha and Re when $Pr = 0.1$, $Ra = 10^3$, $\Delta t = 0.25$ under vertically applied magnetic field. (**a**) $Ha = 10$. (**b**) $Ha = 100$

in Fig. 5.13 compared to the flow in Fig. 5.8 for the same Re and Ha since $R_m = 40$ is also present there.

When external magnetic field applies horizontally, the last term in the vorticity equation (5.13) is replaced by $\frac{Ha^2}{Re}\frac{\partial v}{\partial x}$. Figures 5.15 and 5.16 show Re and Ra effects on the flow and the temperature for this case. When compared to Figs. 5.13 and 5.14, respectively, significant effect is only seen in the flow, as multiple counter flows when Re or Ra increases. Vorticity contours and isolines are not affected much with the direction of the magnetic field.

MHD natural convection flow and heat transfer in a laterally heated enclosure with an off-centered partition have been studied in [77] by using differential quadrature method (DQM). Both the DRBEM and DQM have been applied for solving natural convection flow in a cavity under a magnetic field by Alsoy-Akgün and Tezer-Sezgin [78]. Stabilized FEM procedure has been used by Aydın and Tezer-Sezgin [79] for solving natural convection MHD flow in a sinusoidal corrugated enclosure in terms of primitive variables.

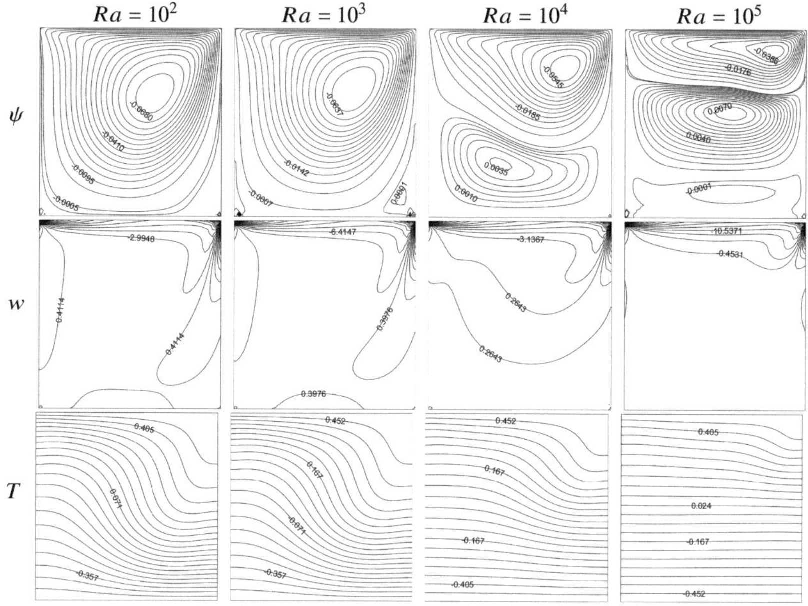

Fig. 5.14 Effects of Ra when $Pr = 0.1$, $Ha = 10$, $Re = 400$, $\Delta t = 0.25$ under vertically applied magnetic field

5.5 Inductionless MHD Duct Flow with Electric Potential

For liquid metal MHD flows under fusion blanket conditions, the following assumptions are appropriate: The fluid is incompressible, and the induced magnetic field is negligible compared to the imposed field B_0. Also, the flow is isothermal, decoupling the energy equation from the momentum equations.

In this problem, the variation of the velocity of the fluid is assumed to be only in the pipe-axis direction (z-axis) as $U = (0, 0, u_z)$, and external magnetic field applies in plane with an angle γ made with the x-axis as $B = (B_0 \cos\gamma, B_0 \sin\gamma, 0)$ giving Eqs. (1.52) derived in Sect. 1.2.6 of Chap. 1, [10].

$$\nabla^2 u_z - Ha^2 u_z = -1 + Ha^2(\sin\gamma \frac{\partial \phi}{\partial x} - \cos\gamma \frac{\partial \phi}{\partial y})$$

$$\nabla^2 \phi = -\sin\gamma \frac{\partial u_z}{\partial x} + \cos\gamma \frac{\partial u_z}{\partial y}\,. \tag{5.15}$$

The region is taken as the square cavity $\{(x, y)| -1 \leq x \leq 1,\ -1 \leq y \leq 1\}$ with the no-slip wall conditions for the velocity

$$u_z(x, \pm 1) = u_z(\pm 1, y) = 0$$

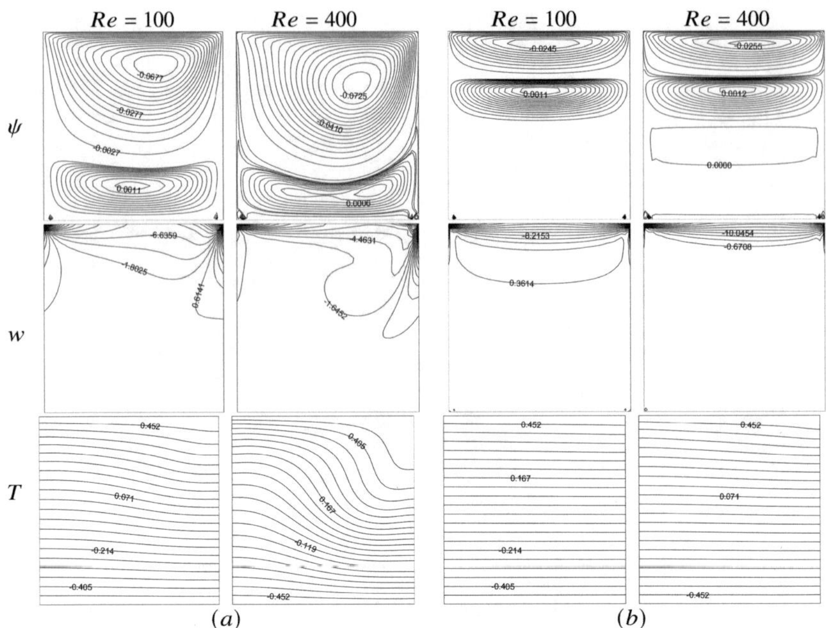

Fig. 5.15 Effect of Ha and Re when $Pr = 0.1$, $Ra = 10^3$, $\Delta t = 0.25$ under horizontally applied magnetic field. (**a**) $Ha = 10$. (**b**) $Ha = 100$

and the fluid/wall interface conditions on the electric potential

$$\pm\frac{\partial \phi}{\partial y}(x, \pm 1) = c\frac{\partial^2 \phi}{\partial x^2}(x, \pm 1)$$

$$\pm\frac{\partial \phi}{\partial x}(\pm 1, y) = c\frac{\partial^2 \phi}{\partial y^2}(\pm 1, y),$$

where c is the wall conduction ratio, that is, c is the measure of the conductance of the wall compared to that of the fluid. The DRBEM discretized equations for Eq. (5.15) become

$$\begin{aligned}(\mathbf{H} - Ha^2\mathbf{C})u_z - \mathbf{G}u_{z_q} &= \mathbf{C}\{-\mathbf{1}\} + Ha^2\mathbf{C}(\sin\gamma \mathbf{D_x} - \cos\gamma \mathbf{D_y})\phi \\ \mathbf{H}\phi - \mathbf{G}\phi_q &= \mathbf{C}(-\sin\gamma \mathbf{D_x} + \cos\gamma \mathbf{D_y})u_z\,,\end{aligned} \tag{5.16}$$

where $\{-\mathbf{1}\}$ denotes the constant vector with components -1 and the matrices $\mathbf{H}$, $\mathbf{G}$, $\mathbf{C}$, $\mathbf{D_x}$, and $\mathbf{D_y}$ are as defined in Sect. 5.2. The equations given in (5.16) for the velocity and the electric potential can be rewritten in the coupled matrix–vector equation form as follows:

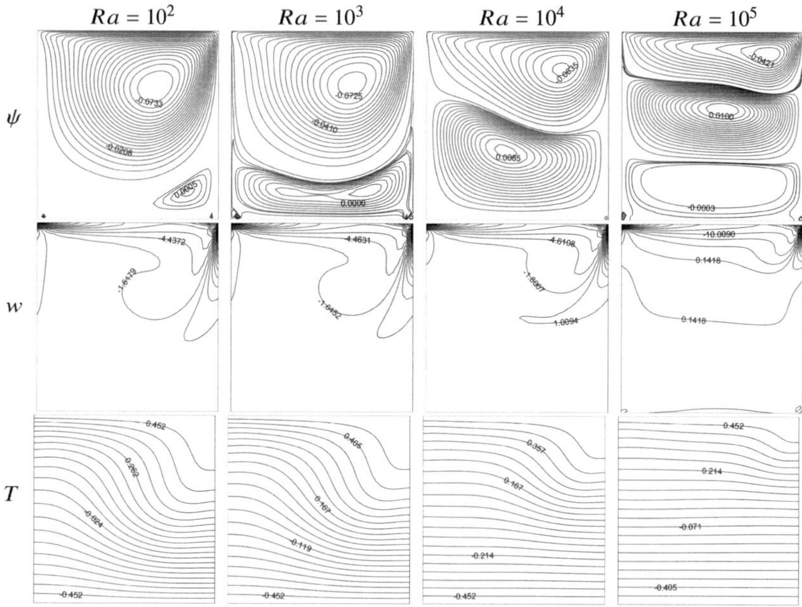

Fig. 5.16 Effects of Ra when $Pr = 0.1$, $Ha = 10$, $Re = 400$, $\Delta t = 0.25$ under horizontally applied magnetic field

$$\begin{bmatrix} \mathbf{H} - Ha^2\mathbf{C} & -Ha^2\mathbf{C}(\sin\gamma\mathbf{D_x} + \cos\gamma\mathbf{D_y}) \\ -\mathbf{C}(-\sin\gamma\mathbf{D_x} + \cos\gamma\mathbf{D_y}) & \mathbf{H} \end{bmatrix} \begin{bmatrix} u_z \\ \phi \end{bmatrix}$$
$$= \begin{bmatrix} \mathbf{G} & \mathbf{0} \\ \mathbf{0} & \mathbf{G} \end{bmatrix} \begin{bmatrix} u_{zq} \\ \phi_q \end{bmatrix} - \begin{bmatrix} \mathbf{C}\{\mathbf{1}\} \\ \mathbf{0} \end{bmatrix}. \tag{5.17}$$

The solution is obtained after the insertion of the boundary conditions and solving the system (5.17) iteratively according to a preassigned relative tolerance $\epsilon = 10^{-4}$ between two successive iterations. The derivative boundary conditions for the electric potential are approximated by using the coordinate matrix **F**.

Results are shown for two-dimensional MHD flow with electric potential in a square duct at different Hartmann numbers and wall conductivities. The velocity u_z depends on the viscosity implicitly through the relations (1.34). As can be seen, the Hartmann and Reynolds numbers are proportional to each other through the viscosity, which implies that the variations of Hartmann and Reynolds numbers have the same effect on the velocity and electrical potential, although Reynolds number does not appear explicitly in Eq. (5.15). The wall conductivities are responsible for the pressure losses and the structure of the flows.

In addition to the influence of the magnetic field (strength and direction), we study the effect of the wall conductance ratios on the flow. The wall conductance ratios of the four walls (c_t (top), c_b (bottom), c_l (left), c_r (right)) can be chosen independently of one another. Hunt [52] gives an analytical solution of the problem in the form of an infinite series for two special cases:

(i) $c_t = c_b \to \infty$ and $c_l = c_r$ are arbitrary.
(ii) $c_t = c_b$ are arbitrary and $c_l = c_r = 0$.

The numerical simulations are performed to analyze the effect of Ha on the velocity and electric potential under the vertically applied magnetic field ($\gamma = \pi/2$) by assuming the nonconducting side walls ($c = 0$ along $x = \pm 1$) and variably conducting Hartmann walls ($c = 0.1$ along $y = \pm 1$) under no-slip boundary conditions. The contour plots of the velocity and electric potential as well as the velocity level curves are presented in Fig. 5.17. The most distinct feature is the side layers where the velocities can exceed those at the center by order of magnitude, as Ha increases. Electric potential slightly clusters through the side walls.

Figure 5.18 shows that both the flow and electric potential are aligned in the direction of externally applied magnetic field. The effect of Ha is the same as in all MHD duct flow cases, which is the flattening tendency and formation of boundary layers as Ha increases. The electric potential is not affected much as behavior change with an increase in Ha but flow concentrates in front of side walls in terms of two loops with a lowered flow level along the left wall due to $c_l = 0.1$ as can be seen in Fig. 5.19.

This same inductionless MHD duct flow problem with electric potential has also been solved by using the DRBEM for several cases of wall conductivities in [80]. These are the cases of nonconducting duct walls, nonconducting side walls, variably conducting Hartmann walls, and variably conducting walls with the same or different conductivity ratios.

5.6 Inductionless MHD Flow Under the Magnetic Field in the Pipe-Axis Direction

In this section, the inductionless MHD flow in channels of which the equations are given in Sect. 1.2.7 is investigated when the external magnetic field is applied in the pipe-axis direction. The velocity is given as $\mathbf{U} = (u_x, u_y, u_z)$ with u_z in the pipe-axis direction. That is, the nondimensional equations are [11]

$$
\begin{aligned}
\nabla^2\psi &= -w \\
\nabla^2\phi &= w \\
\frac{1}{N}(u_x\frac{\partial w}{\partial x} + u_y\frac{\partial w}{\partial y}) - \frac{1}{Ha^2}\nabla^2 w &= 0 \\
\frac{1}{N}(u_x\frac{\partial u_z}{\partial x} + u_y\frac{\partial u_z}{\partial y}) - \frac{1}{Ha^2}\nabla^2 u_z &= -\frac{1}{N}\frac{\partial p_z}{\partial z}
\end{aligned}
\qquad (5.18)
$$

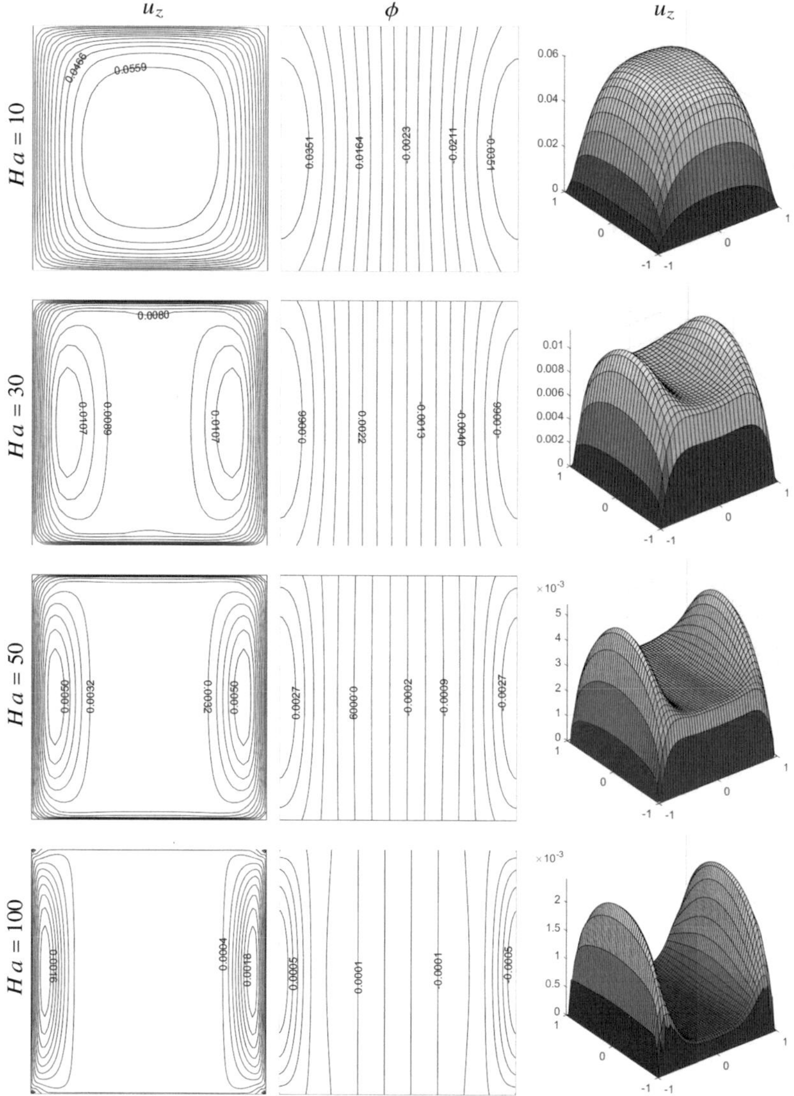

Fig. 5.17 Effect of Ha on the velocity and electric potential for $\gamma = \pi/2$, $c_b = c_t = 0.1$, $c_l = c_r = 0$

as given in Eq. (1.54). On the boundary of the cavity, stream function is a constant due to the known velocity value, electric potential or its normal derivative is zero according to insulated or conducting portions, and the vorticity is computed using DRBEM coordinate matrix from stream function equation.

For the discretization of the boundary, constant elements are used to obtain DRBEM matrix–vector form for Eq. (5.18) as

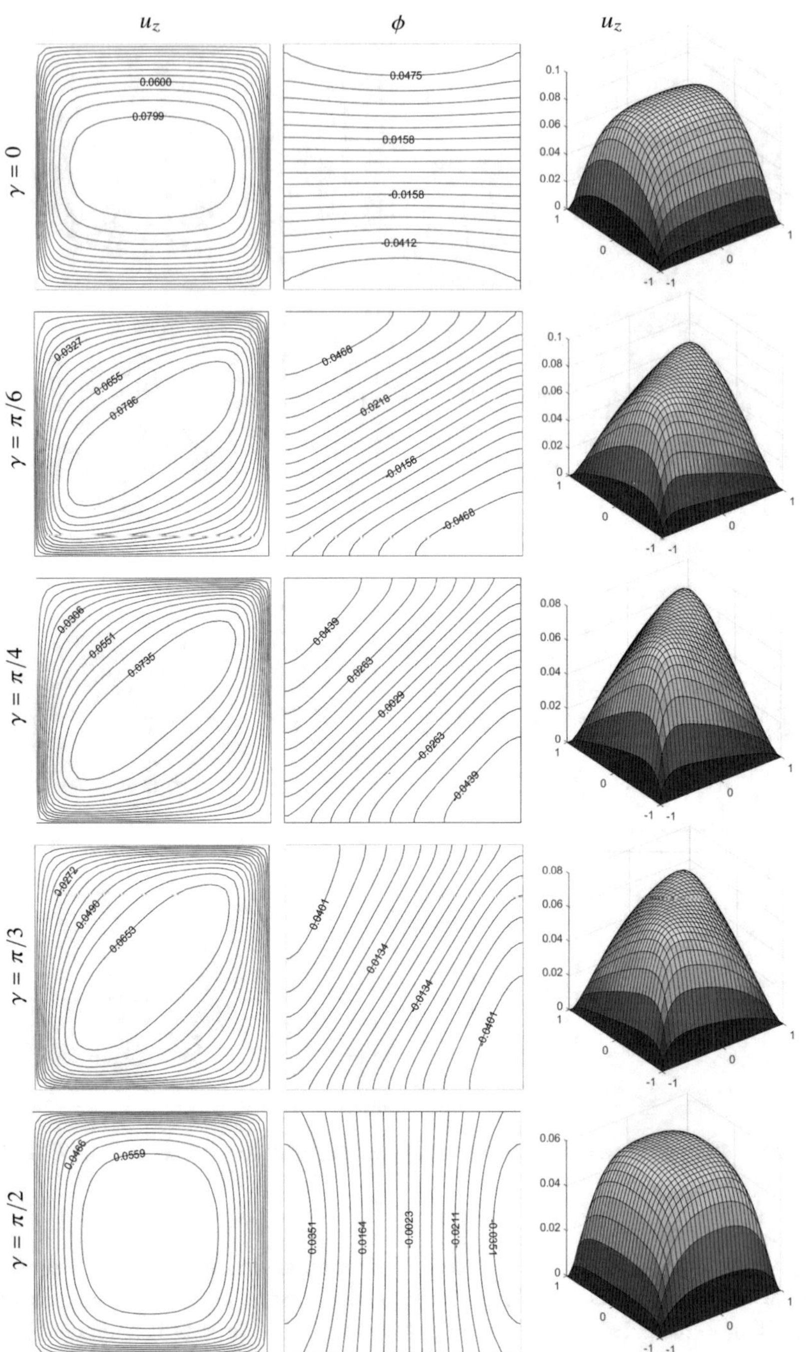

Fig. 5.18 Effect of inclination angle γ on the velocity and electric potential at $Ha = 10$, $c_b = c_t = 0.1$, $c_l = c_r = 0$

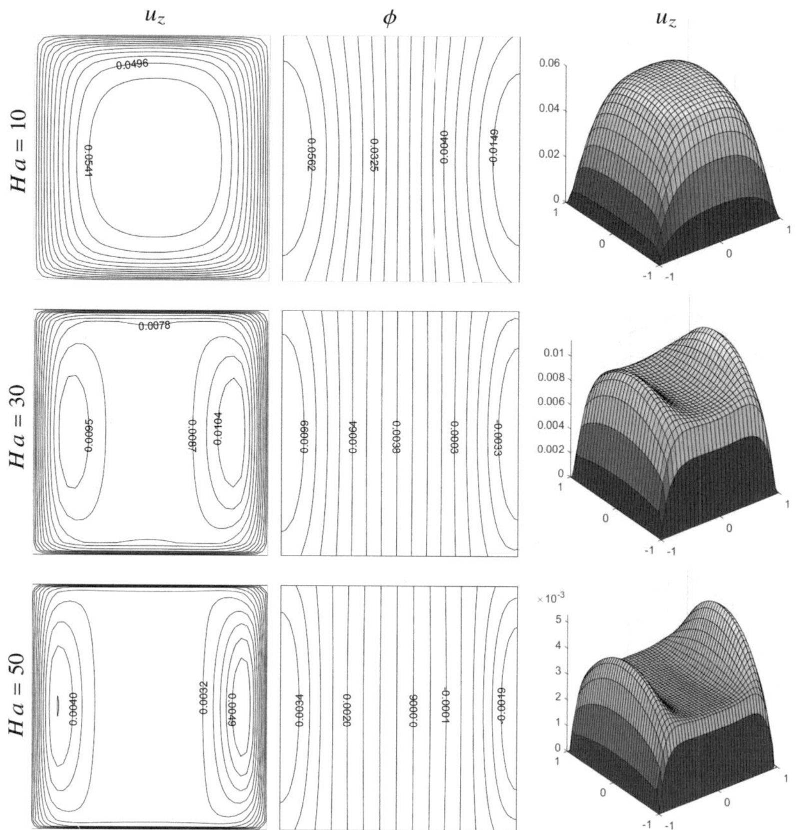

Fig. 5.19 Effect of Ha on the velocity and electric potential for $\gamma = \pi/2$, $c_b = c_t = c_l = 0.1$, $c_r = 0.005$

$$
\begin{aligned}
&\mathbf{H}\psi - \mathbf{G}\psi_q = -\mathbf{C}w \\
&\mathbf{H}\phi - \mathbf{G}\phi_q = \mathbf{C}w \\
&\frac{N}{Ha^2}\left(\mathbf{H} - \mathbf{CK}\right)w - \frac{N}{Ha^2}\mathbf{G}w_q = 0 \\
&\frac{N}{Ha^2}(\mathbf{H} - \mathbf{CK})u_z - \frac{N}{Ha^2}\mathbf{G}u_{zq} = \mathbf{C}\frac{\partial P_z}{\partial z}\,.
\end{aligned}
\tag{5.19}
$$

It is noticed that the velocity and the vorticity equations are solved independently. However, the stream function and velocity potential depend on the vorticity solution. The problem geometry is taken as the lid-driven cavity that is the cross section of the pipe where the top layer is moving in the positive x-direction as in Sect. 5.2.

However, in this problem the external magnetic field in the pipe-axis direction $B = (0, 0, B_0)$ is applied perpendicular to cavity (duct), and it generates electric potential interacting with the electrically conducting fluid in the pipe. Fluid moves with the movement of the lid and the constant pressure gradient $\dfrac{\partial P_z}{\partial z} = -8000$ opposite to pipe-axis direction. In the simulation of the flow and electric potential, $N = 120$ and $L = 900$ constant boundary elements and interior nodes, respectively, are taken for increasing values of Hartmann number Ha, keeping Stuart number $N = 16$ fixed.

In Fig. 5.20 we present streamlines, equivorticity, and equipotential lines in the case of electrically conducting pipe wall ($\phi = 0$) for Hartmann number values $Ha = 20, 100, 200$ that correspond to Reynolds numbers $Re = 25, 625, 2500$, respectively, since $Ha^2 = N\,Re$. It is observed that an increase in the strength of the applied magnetic field (increase in Ha) causes the primary vortex of streamlines to move through the center of the cavity. Recirculations appear at the lower corners and finally at the left upper corner with further increase in Ha and the movement of the lid to the right. Vorticity moves away from the cavity center toward the walls indicating strong vorticity gradients. The fluid begins to rotate with a constant angular velocity, and it flows creating boundary layers near the top and right walls through the upper right corner. Electric potential has the same pattern and magnitudes of streamlines since $\nabla^2\phi = w$, $\nabla^2\psi = -w$, and both ψ and ϕ are zero for this case on the cavity walls. On the other hand, Fig. 5.20 displays the

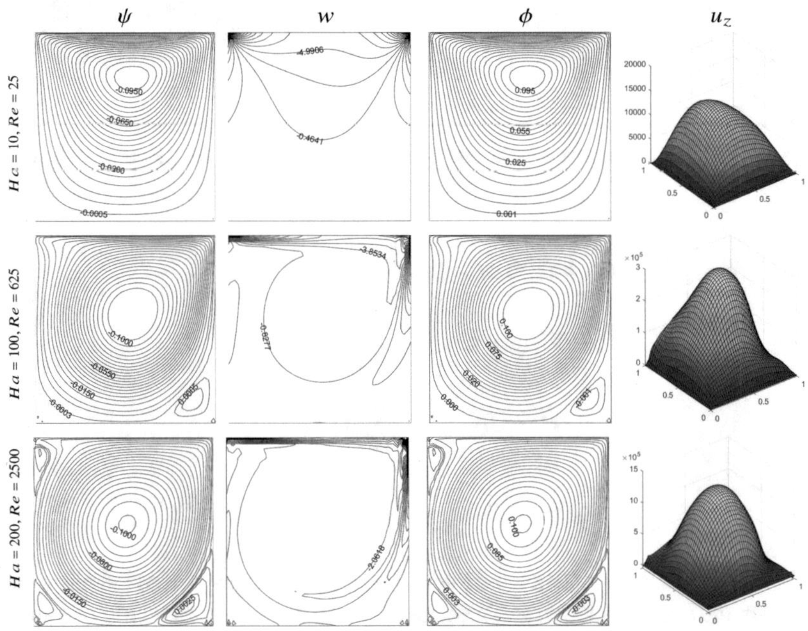

Fig. 5.20 Effect of Ha on ψ, w, ϕ, and u_z when $\phi = 0$ on the walls

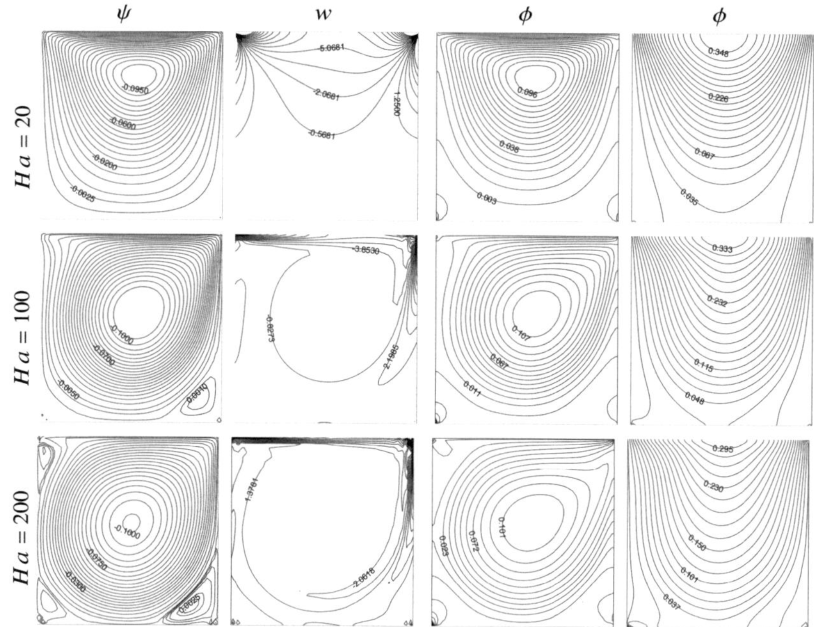

Fig. 5.21 Effect of Ha on ψ, w, and ϕ when $\frac{\partial \phi}{\partial n}|_{x=0,1} = 0$, $\phi|_{y=0,1} = 0$ (third column) and $\frac{\partial \phi}{\partial n}|_{y=0,1} = 0$, $\phi|_{x=0,1} = 0$ (fourth column)

increase in the magnitude of the pipe-axis velocity u_z with an increase in Ha when $\frac{\partial P_z}{\partial z} = -8000$. The damping in the magnitude of u_z is seen close to the moving lid as Ha increases ($Ha = 20, 100, 200$).

When the cavity walls are partly insulated and partly conducting, electric potential leaves the behavior of the flow and obeys boundary conditions on the walls for small values of Ha. It is seen from Fig. 5.21 that insulated vertical walls force the potential first to touch these walls and then both the increased magnetic intensity and moving lid cause it to regain the flow behavior. On the other hand, insulated top and bottom walls give completely different patterns from the flow as traveling electric waves from the bottom to the top. Increasing Hartmann number does not change this behavior much but tends to concentrate through the upper right corner. Both of the cases of electrically conducting and partly insulated–partly conducting cavity walls have been considered in [11]. The behaviors of stream function, vorticity, electric potential, and pipe-axis velocity are in very well agreement with the results obtained here.

In this chapter, the MHD duct flow problems for which the fundamental solution to the whole differential equation is not available, are collected and solved by the use of DRBEM. The fundamental solution of the Laplace equation has been used keeping all the other terms as inhomogeneity. These are unidirectional unsteady

MHD duct flow, MHD flow and heat transfer with magnetic induction, full MHD equations with energy equation, inductionless Buoyancy driven MHD flow, and inductionless MHD flow with electric potential in either plane or pipe-axis direction. The solution of each problem is simulated and discussed for several values of problem parameters to see their effects on the flow, induced current, and electric potential.

References

1. U. Müller, L. Bühler, *Magnetofluiddynamics in Channels and Containers* (Springer, Berlin, 2001)
2. C.A. Brebbia, J. Dominguez, *Boundary Elements: An Introductory Course* (Computational Mechanics Publications, Southampton and McGraw Hill, NewYork, 1992)
3. P. Partridge, C. Brebbia, L. Wrobel, *The Dual Reciprocity Boundary Element Method* (Computational Mechanics Publications, Southampton, Boston, 1992)
4. J. Hartmann, F. Lazarus, *Hg-dynamics II : Experimental Investigations on the Flow of Mercury in a Homogeneous Magnetic Field.* Mathematisk-fysiske meddelelser XV, vol. 7 (Kgl. Danske Viden-Skabernes Selskab, Copenhagen, 1937)
5. P.A. Davidson, *An Introduction to Magnetohydrodynamics* (Cambridge University Press, Cambridge, 2001)
6. R. Moreau, *Magnetohydrodynamics* (Kluwer Academic Publishers, Dardrecht, 1990)
7. L. Dragoş, *Magneto-Fluid Dynamics* (Abacus Press, London, 1975)
8. B. Pekmen, M. Tezer-Sezgin, Int. J. Heat Mass Trans. **71**, 172 (2014)
9. P. Roberts, *An Introduction to Magnetohydrodynamics* (American Elsevier, Amsterdam, 1967)
10. A. Sterl, J. Fluid Mech. **216**, 161 (1990)
11. M. Tezer-Sezgin, C. Bozkaya, in ed. by B. Karasozen, M. Manguoglu, M. Tezer-Sezgin, S. Goktepe, O. Ugur, *Numerical Mathematics and Advanced Applications (ENUMATH 2015).* Lecture Notes in Computational Science and Engineering, vol. 112 (2016), pp. 3–11
12. M. Sezgin, Some mixed boundary value problems in magneto-hydrodynamics. PhD Dissertation, University of Calgary, 1983
13. J.A. Shercliff, J. Fluid Mech. **1**, 644 (1956)
14. S. Gilbert, G. Fix, *An Analysis of the Finite Element Method* (Wellesley-Cambridge Press, Wellesley, 2008)
15. J. Reddy, *An Introduction to Finite Element Method* (McGraw-Hill Education, New York, 2005)
16. C.A. Felippa, *Introduction to Finite Element Methods* (WordPress.com, 2004)
17. C.A. Brebbia, *The Boundary Element Method for Engineers* (Pentech Press, London, 1978)
18. C.A. Brebbia, *Recent Advances in Boundary Element Methods* (Pentech Press, London, 1978)
19. C.A. Brebbia, S. Walker, *Boundary Element Techniques in Engineering* (Newness-Butterworths, London, 1979)
20. C. Brebbia, J. Telles, L. Wrobel, *Boundary Element Techniques* (Springer, Berlin, 1984)
21. J.T. Katsikadelis, *Boundary Element Methods: Theory and Applications* (Elsevier, Oxford, 2002)

M. Tezer-Sezgin, C. Bozkaya, *Boundary Element Method for Magnetohydrodynamic Flow*, Surveys and Tutorials in the Applied Mathematical Sciences 14, https://doi.org/10.1007/978-3-031-58353-7

22. C. Pozrikidis, *A Practical Guide to Boundary Element Methods with the Software Library BEMLIB* (Chapmann Hall/CRC Press, Boca Raton, 2002)
23. F. Paris, J. Canas, *Boundary Element Method: Fundamentals and Applications* (Oxford University Press, Oxford, 1997)
24. G.S. Gibson, *Boundary Element Fundamentals: Basic Concepts and Recent Developments in the Poisson Equation* (Computational Mechanics Publications, Southampton, 1987)
25. P.K. Banerjee, *The Boundary Element Methods in Engineering* (McGraw-Hill College, New York, 1994)
26. W.S. Hall, *The Boundary Element Method* (Springer Dordrecht, Berlin, 1994)
27. F. Hartmann, *Introduction to Boundary Elements: Theory and Applications* (Springer, Berlin, 1989)
28. A.A. Becker, *IThe Boundary Element Method in Engineering: A Complete Course* (McGraw-Hill, New York, 1992)
29. G. Chen, J. Zhou, *Boundary Element Methods* (Academic, Cambridge, 1992)
30. P.K. Banerjee, R. Butterfield, *Boundary Element Methods in Engineering Science* (McGraw-Hill, New York, 1981)
31. G. Beer, M.I. Smith, C. Duenser, *The Boundary Element Method with Programming* (Springer, Wien, 2008)
32. S. Rjasanow, O. Steinbach, *The Fast Solution of Boundary Integral Equations* (Springer, Berlin, 2007)
33. D. Nardini, C.A. Brebbia, *A New Approach to Free Vibration Analysis Using Boundary Elements* (Computational Mechanics Publications and Springer, Berlin, 1982)
34. C.F. Loeffler, W.J. Mansur, *Boundary Elements X: Dual Reciprocity Boundary Element Formulation for Potential Problems in Infinite Domains* (Computational Mechanics Publication/Springer, Southampton/Berlin, 1988)
35. R. Zheng, C. Coleman, N. Phan-Thien, Comput. Mech. **7**, 279 (1991)
36. S. Gümgüm, The dual reciprocity boundary element method solution of fluid flow problems. PhD Dissertation, Middle East Technical University, 2010
37. C. Bozkaya, Boundary element method solution of initial and boundary value problems in fluid dynamics and magnetohydrodynamics. PhD Dissertation, Middle East Technical University, 2008
38. K.M. Singh, M.S. Kalra, *Numerical Methods in Thermal Problems*, vol. I (Pineridge Press, Swansea, 1989), p. Part I
39. A. Lahrmann, C. Haberland, *Numerical Methods in Thermal Problems*, vol. II (Pineridge Press, Swansea, 1991), p. Part I
40. C.P. Rahaim, A.J. Kassab, Eng. Analy. Boundary Elements **18**(4), 265 (1996)
41. W.J. Mansur, F.C. Araújo, J.E.B. Malaghini, Int. J. Numer. Methods Eng. **33**(9), 1823 (1992)
42. A. Saitoh, A. Kamitani, IEEE Trans. Mag. **40**(2), 1084 (2004)
43. F. de Araújo, E. d'Azevedo, L. Gray, Eng. Analy. Boundary Elements **35**(3), 517 (2011)
44. M. Wathen, C. Greif, D. Schötzau, SIAM J. Sci. Comput. **39**(6), A2993 (2017)
45. C. Bozkaya, M. Tezer-Sezgin, J. Comput. Appl. Math. **203**(1), 125 (2007)
46. B. Singh, J. Lal, Int. J. Numer. Methods Eng. **18**(7), 1104 (1982)
47. Z. Demendy, T. Nagy, Acta Mechanica **123**, 135 (1997)
48. A. Nesliturk, M. Tezer-Sezgin, J. Comput. Appl. Math. **192**(2), 339 (2006)
49. X. Cai, H. Qiang, S. Dong, J. Lu, D. Wang, *Proceedings of the 2018 2nd International Conference on Applied Mathematics, Modelling and Statistics Application (AMMSA 2018)* (Atlantis Press, Amsterdam, 2018), pp. 68–71
50. X. Cai, G. Su, S. Qiu, Appl. Math. Comput. **217**(9), 4529 (2011)
51. J.A. Shercliff, Proc. Camb. Phil. Soc. **49**, 136 (1953)
52. J.C.R. Hunt, J. Fluid Mech. **21**(4), 577–590 (1965)
53. C.C. Chang, T.S. Lundgren, Zeitschrift für Angewandte Mathematik und Physik ZAMP **12**(2), 100 (1961)
54. M. Tezer-Sezgin, S. Dost, Appl. Math. Modell. **18**(8), 429 (1994)
55. R.R. Gold, J. Fluid Mech. **13**(4), 505–512 (1962)

56. M. Tezer-Sezgin, S.H. Aydın, Eng. Analy. Boundary Elements **30**(5), 411 (2006)
57. C. Bozkaya, M. Tezer-Sezgin, Eur. J. Comput. Mech. **26**(4), 394 (2017)
58. G.A. Grinberg, PMM **25**(6), 1024 (1961)
59. M. Tezer-Sezgin, Int. J. Numer. Methods Fluids **18**(10), 937 (1994)
60. M. Tezer-Sezgin, C. Bozkaya, Comput. Mech. **41**(6), 769 (2007)
61. M. Sezgin, Int. J. Numer. Methods Fluids **7**(7), 697 (1987)
62. I. Butsenicks, E. Shcherbinin, Magn. Girodinam **2**, 35 (1976)
63. M. Sezgin, Int. J. Numer. Methods Fluids **8**(6), 705 (1988)
64. C. Bozkaya, M. Tezer-Sezgin, Comput. Fluids **66**, 177 (2012)
65. C. Bozkaya, M. Tezer-Sezgin, Int. J. Numer. Methods Fluids **70**(3), 300 (2012)
66. M. Sezgin, Int. J. Numer. Methods Fluids **8**(7), 743 (1988)
67. M. Tezer-Sezgin, C. Bozkaya, J. Comput. Appl. Math. **225**(2), 510 (2009)
68. C. Bozkaya, M. Tezer-Sezgin, Eng. Analy. Boundary Elements **36**(4), 591 (2012)
69. J.C.R. Hunt, W.E. Williams, J. Fluid Mech. **31**(4), 705–722 (1968)
70. M. Tezer-sezgin, M. Gürbüz, J. Balıkesir Univer. Instit. Sci. Technol. **20**(3), 53 (2018). https://doi.org/10.25092/baunfbed.476597
71. M. Dehghan, D. Mirzaei, Comput. Phys. Commun. **180**(9), 1458 (2009)
72. N. Bozkaya, M. Tezer-Sezgin, Int. J. Numer. Methods Fluids **56**(11), 1969 (2008)
73. C. Bozkaya, M. Tezer-Sezgin, Int. J. Numer. Methods Fluids **51**(5), 567 (2006)
74. E. Ebren Kaya, M. Tezer-Sezgin, Eng. Analy. Boundary Elements **117**, 242 (2020)
75. B. Pekmen, DRBEM applications in fluid dynamics problems and DQM solutions of hyperbolic equations. PhD Dissertation, Middle East Technical University, 2014
76. B. Pekmen, M. Tezer-Sezgin, CMES-Comput. Model. Eng. Sci. **105**(3), 183 (2015)
77. K. Kahveci, S. Öztuna, Eur. J. Mech.-B/Fluids **28**(6), 744 (2009)
78. N. Alsoy-Akgün, M. Tezer-Sezgin, Progr. Comput. Fluid Dyn. **13**(5), 270 (2013)
79. S.H. Aydin, M. Tezer-Sezgin, Int. J. Comput. Math. **97**(1–2), 420 (2020)
80. E. Ebren Kaya, BEM solutions of magnetohydrodynamic flow equations under the time and axial-dependent magnetic field. Ph.D. Thesis, Middle East Technical University, 2021

Index

M. Tezer-Sezgin, C. Bozkaya, *Boundary Element Method for Magnetohydrodynamic Flow*, Surveys and Tutorials in the Applied Mathematical Sciences 14, https://doi.org/10.1007/978-3-031-58353-7

Printed in the United States
by Baker & Taylor Publisher Services